T0255991

Textile Science and Clothing Technology

Series editor

Subramanian Senthilkannan Muthu, Kowloon, Hong Kong

More information about this series at http://www.springer.com/series/13111

Subramanian Senthilkannan Muthu
Editor

Sustainable Innovations in Apparel Production

 Springer

Editor
Subramanian Senthilkannan Muthu
Kowloon
Hong Kong

ISSN 2197-9863   ISSN 2197-9871 (electronic)
Textile Science and Clothing Technology
ISBN 978-981-13-4193-9   ISBN 978-981-10-8591-8 (eBook)
https://doi.org/10.1007/978-981-10-8591-8

Printed on acid-free paper

This Springer imprint is published by the registered company Springer Nature Singapore Pte Ltd. part of Springer Nature
The registered company address is: 152 Beach Road, #21-01/04 Gateway East, Singapore 189721, Singapore

This book is dedicated to:
The lotus feet of my beloved
Lord Pazhaniandavar
My beloved late Father
My beloved Mother
My beloved Wife Karpagam and
Daughters—Anu and Karthika
My beloved Brother
Last but not least
To everyone working in the global apparel
sector to make it SUSTAINABLE

Contents

Laser-Based Apparel Production

P. Senthil Kumar and S. Suganya

Abstract LASER (Light Amplification by Stimulated Emission of Radiation) technology is being largely used in apparel industry for cutting, patterning garments, designer neckties, and denim fading with 3D body scanning and engraving leather since nineteenth century. Laser cut design tends to be reserved for haute couture designs and reduced low cost, flexibility, and anti-counterfeiting to produce apparel in ready-to-wear collections. Laser light is a form of electromagnetic radiation used to cleave various materials with high accuracy in cutting, sealing fabric edges in order to prevent fraying. The change in energy states within the atoms of certain materials leads to produce light by laser. And that has few basic characteristics, namely intensity, coherency, monochromaticity, and collimation. These are helpful to distinguish laser light from natural light. Generally, laser beams are narrow, travel in parallel lines, but do not spread out or diverge as light from most normal sources. Therefore, using laser cuts without any pressure on the fabric is meant for no extra energy requirement other than laser. It tends to no unintended marks left on the fabric especially in silk and lace. Adopting high-energy laser cuts material by melting, burning, or vaporizing it. Most significantly, laser beam decomposes dye, resulting in producing vapors followed by venting them away from garment. This is how denim fading works. While scanning the universal barcodes to identify products such as apparels, fashion accessories, the following lasers are used such as CO_2 laser, neodymium (Nd) laser, and neodymium yttrium-aluminum-garnet (Nd-YAG) lasers. They use precise concentrated beam of light. CO_2 laser is a gas laser, producing an infrared light to absorb by organic material. Solid-state lasers such as Nd and Nd-YAG lasers, on the other hand, rely on a crystal to create light beam. Yet, it is hard to reproduce in an exact way. Hence, laser cut makes each ideal task to create an identical design; many countries are unaware of this technology. But the laser cut clothes are shell out for a lot of cash. However, safety issues and gases used in laser apparels must be replenished to meet

P. Senthil Kumar (✉) · S. Suganya
Department of Chemical Engineering, SSN College of Engineering, Chennai 603110, India
e-mail: senthilchem8582@gmail.com; senthilkumarp@ssn.edu.in

© Springer Nature Singapore Pte Ltd. 2018
S. S. Muthu (ed.), *Sustainable Innovations in Apparel Production*,
Textile Science and Clothing Technology,
https://doi.org/10.1007/978-981-10-8591-8_1

multi-fiber agreement regime to make textile products more safe, clean, and competitive. This chapter focuses on laser technology in the apparel production and their potential hazards in health-related concerns.

Keywords Laser cut design · Engraving · Scanning · 3D · Hazards
Light beam

1 Introduction

For many decades, LASER technology is being used in apparel industry. Recently, laser technological developments are gripping textile field to improve the product and supply chain more efficient. Laser plays major role in cutting, patterning, welding and designing, scanning, fading, and engraving [1]. Laser light is an energy source, a form of electromagnetic radiation which produces light by changing energy states within the atoms of certain materials. Light emitted from laser falls on following characteristics, namely intensity, coherency, monochromaticity, and collimation. Though laser beams are not divergent, their intensity and power can be precisely controlled. Furthermore, laser can focus to a desired object at specific angle according to its application [2]. Similarly, laser passes on denim to decompose dye and vaporize that further results in fading. In particular, helium–neon lasers are used to scan the universal barcodes to identify products such as apparels, fashion accessories in retail stores. It also strengthens the security policy for goods sold in market, preventing from duplicating. To overcome such economical issue, digital storage of goods is encouraged rather than physical patterns storage. Digital artwork storage can be later converted into physical samples using lasers.

Denim-engraving techniques have been developing for value addition purpose using laser which can be a substitute for traditional methods (Denim-distressing), whereby laser-based methods offer high degree of sophistication to the denim segment compared to non-laser methods. Laser is feasible at the extent of cutting flexible objects, i.e., fabric to rigid metal. To achieve this, industrialists are looking forward to laser equipment for its advantages in accuracy, efficiency, simplicity, and the scope of automation. Traditional cutting tools for cutting garments mostly damage the quality of objects especially on exquisite materials as the cutting force [3] is applied to band blades, disks, and reciprocating knives. In other words, it leads to inaccurate cutting.

Traditional methods also require man power such as operator to pay full attention during cutting to trade-off a trade-off between two incompatible features, i.e., cutting speed and its accuracy. Indeed, limitations in traditional methods include intricacy of the cut components, tool longevity, and machine downtime during tool servicing. While employing laser-based tools and devices, aforesaid limitations can be eliminated with improved efficiency and reduced cost. Use of laser in apparel production contributes high speed in cutting and processing, high precision, and simple operation. It forms pavement to cut leather graphics to draw any desirable clothing model. During this, laser beam is collimated to focus on a fine dot for precise cutting. This also utilizes least size of the garment with high efficiency and exquisite tailoring than the non-laser-cutting methods by spectrum. Laser technology aims to reduce cost in processing, flexibility in product development, and anti-counterfeiting. In addition, garment industry eliminates tedious handling systems operated by workers in non-laser workstations in order to implement safety guidelines [4].

This chapter aims to provide the major role of laser in design and apparel production. Inevitably, recent advances in laser technology depend on computer programs for geographical design and financial control in all kinds of markets to strengthen communication. It assures to assist speed and accuracy. It also discusses the complexity of the subject and interprets the advances in apparel production for easy and better understanding of students that to develop young industrialists right away in this dynamic industry.

Health-related concerns in developing laser technology and its appliance are discussed. Moreover, product damage or spoilage; no/less consumables; less toxicity during by-product disposal activities while using laser in garment industry are discussed in subsequent sections. Adapting laser equipment is again interesting to develop new products. Earlier laser systems were cumbersome, hard to run, and difficult to maintain, but modern systems are simpler and easy to operate and maintain. Furthermore, gas replenishment in an earlier system is not recommended by experts. Yet, garment manufactures across the globe are producing their products safer and competitive as they sign multi-fiber agreement regime [5].

Some sections are very normative and descriptive based on current research in apparel sizing and fit into computerized pattern, which may integrate young researchers and diverse professionals to stimulate their ideas according to their own interest and specialisms.

2 Laser Technology

2.1 History of Laser

A laser system generates monochromatic, i.e., single color, coherent photons, i.e., emissions with a uniform constant frequency and wavelength in a low-divergent beam [6]. A focus area of laser in fabric cutting is 0.1 mm in diameter and energy of 75 W. It is evident for laser processing being more energetic than sunlight. Laser was first demonstrated in California, 1960. The Bell Laboratories, New Jersey, have begun using CO_2 lasers in late 1964. During the 1980s, laser-based fabricated products were dominated the apparel industry [7].

2.2   Laser Technology

Laser is a coherent, focused beam of photons, traveling along with same wavelength. It is an electromagnetic energy; all the waves have same phase and frequency. Laser is an electromagnetic radiation like ordinary light and infrared but changes in energy states within atoms are responsible for showing off light. At the same time, when atoms are promoted from ground state to excited state, i.e., high energy states, it emits laser in the form of light, called as 'stimulated emission.' Notably, the color of laser is fixed by its wavelength which is in nanoscale.

2.3   Types of Laser

Types of laser are commonly distinguished by fundamental characteristics. Different kinds of lasers are achievable from different kinds of atom excitements through different mediums such as solid, gas, liquid.

Solid-state lasers are mostly passed through medium such as ruby rod or any solid material in crystalline form. This solid can be doped and replaced with ions of impurities for enabling energy levels to produce laser light. Precise frequency is possible when high-powered beams are produced typically in very brief pulses.

Gas lasers are powerful and efficient and, by contrast, produce continuous bright beams using noble gases like CO_2 as their medium. It is pumped by electricity.

Liquid dye lasers utilize a solution as medium in which organic dye molecules are pumped by an arc lamp. It produces broader band of light frequencies. It is advantageous over solid-state and gas lasers in 'tuning' to produce different frequencies [8, 9].

Semiconductor lasers are cheap and tiny and look like chip-based devices. It dually works between a conventional light-emitting diode (LED) and a traditional laser. It is also known as laser diodes or diode lasers because they generate coherent and monochromatic light by joining missing electrons and holes together.

Fiber lasers are one kind, working inside optical fibers. In which, a doped fiber-optic cable acts as the amplifying medium. Such lasers are powerful, efficient, and reliable with ease of piping laser light.

A laser beam containing high energy concentration per unit area can be of very intense only with 1–2 mm of beam diameter. Unlike all lasers, few of them are not powerful, in spite of high intensity. To say, ordinary light is incoherent, is generated at different times, and propagates in all possible directions randomly. However, laser is coherent and passes through a fine dot in which they propagate [7]. The coherency nature of laser relates to the monochromatic results in the formation of highly collimated laser. As all the waves propagate in same phase in parallel lines, there is almost no divergence as observed in light. This property of laser produces high intensity even after traveling a long distance. By using optical lens to focus a beam, energy concentration of the beam can be increased.

For such good reasons, laser engraving, laser cutting, and laser marking have been advised over textile industry by storm; with no wastage, wear, or loss of quality, laser technology requires no material fixation because of force-free and contact-free processing steps. Say an emphatic 'no' to frayed fabrics, lint, and unclean cuts. Simple production is dependent on PC design programs and advanced laser machines. Driving force of textile-filled advancement is precision, cost-effectiveness, best technological inputs, speedy deliveries, contact-free and force-free processes, etc. Synthetic fibers are friendlier for ease of cutting and styling by laser technology, in compensations, readily adopted [10].

3 Applications of Laser in Apparel Production

Laser in fabric production unit in the apparel industries is summarized.

3.1 Laser Technology in Digital Colorings

Research into design, drape, and production shows historical development in apparel production over the last 30 years. On the other hand, overcoming remarkable obstacles in modeling, measuring, and predicting fabric drape in a garment is encountered by employing laser-based approaches. The modern techniques involve in scanning, mass customization, 3D design, simulation, 3D virtual prototyping, and Web-based shopping. Designer must consider fast fashion demands throughout the supply chain process. Yet, standardizing computerized color matching is problematic. Therefore, an electric form of digital color fingerprint [11] is used in laser technology for communication purpose and to misunderstandings. The primary concern is design and its simulation in account to fabric development. In extension, fit and size of the completed garment have to be reliable to avoid confusion in sizing systems. Good fit is mantra to avail consumer satisfaction. In practice, empirical data for geographical sizing can reflect in accurate size codes and allowances.

Unfortunately, sometimes, the inaccuracies in sizing can occur due to impractical measurements. Such invasive, time-consuming steps could be avoided by analyzing customers' or on the individual's perception of tight or loose fitting, how comfortable their clothes to be. The 3D body-scanning techniques can resolve fitting issues. It acts as a bridge between suppliers and buyers in enabling designers to communicate their ideas. For example, female flak jacket is designed and commercialized with more specifications. Not only designing, laser technology also speeds up the product development in the fast fashion market. Pattern construction is an art to be integrated into a more continuous process. Digital printing links design with manufacturing, marketing, and distribution. Knitting is a kind of construction method to shape garments using 3D technology. However, textile industries do not yet implement modern sewing machineries other than 2D. Thus, executing 3D machines may restructure the traditional methods, being crucial to bring integration to internal systems and to the supply chain [12].

Adopting laser-based devices is more advantageous in cutting, drawing, designing, and precise clothing model with ease of operation and low-energy consumption. The main objective is to achieve high efficiency and exquisite tailoring better than traditional methods, i.e., manual cutting by spectrum. Advances in laser devices answer such practical questions and satisfy the customer demand and accurate sewability of material. This also provides guidelines for future directions with systematic description [11, 12].

3.2   Fabric Fault Detection

All textile industries focus to produce competitive fabrics, which is bit difficult in terms of productivity and quality. Indeed, competitive fabric is meant for 'free from faults' that relies on the identification fault in fabric. Because most of the garment sector invests in large amount of material input, causing drastic loss in future without being aware of faulty raw material or final product. Therefore, identification of faults can be initialized in the garment production unit. Considering 45–65% manufacturer's profit out of 85% margin is due to the second or off-quality goods. Therefore, detecting faults, preventing them from reoccurring, is important.

Human visual inspection has been followed in most of the developed countries to identify the defects occurred during the production process of a textile material. It is very tedious and requires man power and time-consuming. Moreover, small or minute defects are impossible in visual detection. Such average identification rate in real time is helpless in engraving the quality of fabric. This gap is filled by digital image processing to analyze textured samples by laser.

Significantly, laser-based optical Fourier transform analysis is suggested in order to reduce the wastage or to analyze the repeatability of pattern at regular intervals. Laser passes to the point of fabric, and the diffraction gratings obtained from the periodicity of longitudinal and transverse threads in the fabric are superimposed. In such way, a variety of Fourier lens have been used to produce the diffraction patterns of the fabric. Further, to magnify, the textured sample image is inverted with same focal length. A charge-coupled device (CCD) camera has been used to capture image. Accuracy and early-stage detection of defects in fabrics are achieved by transferring data collected so far and stored in a computer. By comparing the acquired images with the stored images, a fault can be identified while converting an image into binary mode in relation to the data deviated from the standard. The severity of the fault depends upon the amount of deviation from the standard occurs. Early-stage detection is a key factor of quality improvement. The developing and developed countries almost installed automated inspection [13, 14].

The faults can be certainly hole, scratch, stretch, fly yarn, dirty spot, slub, cracked point, color bleeding, etc. If they are not detected at its entry level to the production unit, it can affect the production process massively. This section also offers some aspects of fault detection such as inspection type either visual or automated; fabric types; rate of defect and reproducibility; objective of defect; judgment analysis based on statistics, inspection speed, type of response, and information for exchange.

Machine malfunctions do cause orientation issues along pick direction. They seem to be long and narrow. Machine oil spoilage is often along with the warp that seems to be wide and irregular. Such recognizing system as control agent transforms the captured digital image into adjusted resultant output and operates the automated machine. Once recognizer identified a fault in fabric, immediately trigger the laser beams in order to display the upper offset and the lower offset of the faulty portion. And that area is removed. After cutting the faulty portions of fabric, textile defect recognizer resumes its operation. An automated defect detection and identification system enhance the product quality resulted in improved productivity to meet both customer needs and to reduce the costs associated with off-quality.

Technology upgradation on computer vision methodology and adaptive neural networks combines artificial neural networks with image processing in textile industries. Rather than manual fault detection after a sufficient amount of fabric produced, removed from machine, and then batched into larger rolls, sending back to the inspection frame, optimal solution is recommended as automatic inspection to alert the maintenance personnel when the machine needs attention to prevent production of defects or to change process parameters to prevent damage of materials. Based on the dimension of the faults, the fabric is classified, which is graded in market accordingly. An investment in the automated fabric inspection system is economically attractive as reduction in personnel cost and associated benefits are considered.

3.3 Laser Cutting

Fraying problem associated with conventional knife cutters is resolved after an implementation of laser cutters. A well-finished edges are produced as a result of laser melting and fusing at the edges. Fashion accessories, jewelries, and leather are very common, expecting for precision of cut components. Desired pattern on those is attained by passing laser on the surface of fabric, increasing temperature substantially cutting with the help of vaporization. In case of gas lasers, a cutting head fits with mirrors to reflect the laser beam to the portion of cutting line. A computer is used to audit the cutting portions and to control the entire system.

Inert gases such as (N_2, He) [16] are used to prevent the charring and to remove debris and smoke from the cutting area. Laser beam is very sharp unlike mechanical cutting devices. Automatic single-ply laser cutters are faster (30–40 m/min) than automatic multiple-ply knife cutters (5–12 m/min). However, while cutting multiple plies, knife cutters are faster per garment cut and also cheaper. A perfect sealed edge by laser is clean and tangible. A number of lays or layer recommended for cutting are possible in using laser devices. In case of synthetics, there is a chance of the cut edges to be fused together. Multiple layer cutting using laser comes up with special finish in end product. A choice of single-ply cutting knives or multiple-ply cutting by laser is dependent on the purpose of operation. In addition, laser cutting is used in some areas of home furnishing.

Laser processing is a non-contact, tool-free process. The textiles and fabrics are not even touched during laser cutting. With effect, the material is not warped and melted by the way. The results are clean with perfectly sealed edges.

Advantages

- Cheaper and safer,
- Do not need mechanical action,
- High precision,
- Feasible for edge cutting,
- Simple maintenance for longer duration.

Example

Laser cutting is natural and hence proved by two companies, namely 'Cut Laser Cut' and 'Laser Cut Fabric.' They manufacture their own designs of fabric using lasers. EDITD is the world's leading fashion retailers that offer the laser-cutting trend in haute couture collections for hundreds of brands. The business also ends with the cost spectrum and everything in-between [12].

The features of laser applications include:

- Combination between marking, engraving, and cutting is simple;
- High quality from force-free processing;
- No more fraying problems;
- Clean and lint-free fabric;
- Integrated computer design;
- Cost saving and high working speed;
- Investment in high-quality raw material;
- Minimal waste production;
- Perfect sealed edges and accurate contour cutting;
- Productive and non-contact processing;
- Simple high-quality manufacturing;
- One tool for all shapes;
- No tool wear.

Additionally, synthetic fabrics like plastic (polyester) respond very well to laser processing in a controlled manner. It offers fiber-free sealed edges. Choosing appropriate laser parameters and lens can ignore a brownish discolored cutting edge associated with organic fabrics. Styrene-butadiene rubber (SBR) is suggested by an expert for laser processing other than neoprene-based blends.

3.3.1 Laserflex—Multilayered Film

Laserflex [17] is a high-quality, multilayered film and enriches textile designs and logos with a screen-like finish. Optimal laser settings employ high laser power with high-speed plotter in order to remove dust produced during laser processing.

3.3.2 An Argument Regarding Laser Cutting

An extreme precise cut in fabric is accurate without ever touching the fabric. It indicates that the resulting product is intended by a manufacturing process as possible [18].

Key components

- Precision like if a design is done by hand;
- Faster pace;
- More practical;
- Lower price points.

This kind of manufacturing method makes an argument regarding less to be copies. Though the intricate designs are hard to reproduce, the original pattern can be recreated or inspired by specific cuts. But in textile market, industrial designs get patented to make such unique laser cuts harder to create an identical pattern. Unsurprisingly, laser cut designs are found all over, which means one can should shell out a lot of cash on a laser cut garment. Isn't persuasive? So, to buy the clean clothing, laser cut is the fine option to go and spend money.

3.3.3 Laser in Washing

Laser wash has been integrated from 'water-free technology' which is an exceptional method, using no water or stones to wash the jeans. In which, laser beam is passed across the jeans and washed. Moreover, washing efficiency can create extraordinary jeans and environmentally reduce the impacts to the quality of the products as well the surroundings. The wash is quick; finishing looks nice. The laser washing offers more desirable 'miffed' look to the denim. This water-free technology is an apt replacement for traditional methods that is because most of the denim brands have been adopting this technique. Compared to sandblasting, laser washing has the fastest trend in the denim market, sounds denim go green. Additionally, different patterns certainly letter, dots, or lines and even images can be executed through computer-aided programs [19].

3.3.4 Laser Based Denim Fading

In the nineteenth century, sandblasting, hand sanding, destroying, and grinding have been followed for the purpose of fading of denim which becomes out of fashion now. Laser fading is a dry process trending in modern days with a computerized system, whereas laser beam focusses at the point of fading required. The laser beam degrades any coloring agent faster than the older techniques followed by venting away the vapors produced during the process as the laser beam focuses and impacts a fine surface on the fabric which makes fading faster and accurate. Solid-state-based and gas-based lasers are famous to apply in industry. Their fading rate is proportional to wavelength and power. Laser fading is safer compared to acid washing, environmentalists say.

Mechanism of Laser Fading

A laser beam creates an extensive heat within the focused region of fabric; as such, heat energy is absorbed and melted from solid to liquid phase. Due to the surface tension of the liquid, partially molten liquid tries to move. Rest of the liquid phase is heated up rapidly followed by boiling and releasing vapors. Now, the phase gets changed from liquid to gas. This is how laser works in fading [21] as shown in Fig. 1.

Advantages of laser fading:

1. Any portion of fabric can be washed.
2. Fading at the place of metal buttons is possible.
3. The process is faster and accurate.
4. No loss in material quality or strength.
5. No water or chemical usage.
6. Low hairiness and less man power.
7. Environment-friendly.

Disadvantages of laser fading:

1. Small- or medium-scale industries cannot invest huge initial investment;
2. Operator skills are compulsory;

Fig. 1 Mechanism of laser fading

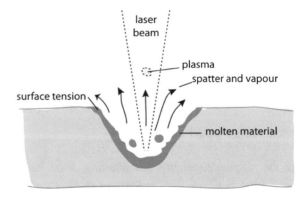

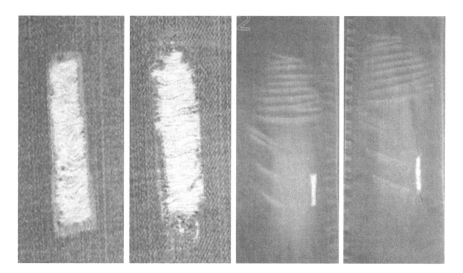

Fig. 2 Laser faded denim

3. To produce natural look after laser washing is difficult;
4. Working in laser is health hazardous;
5. At particular circumstances, maintenance and servicing of the laser system are troublesome.

Laser fading is compared with the manual system as shown in Fig. 2 for the betterment of quality.

Laser Safety

Laser safety is important for labors who are working in laser control unit. The international safety regulations (IEC 60825-1) [22] control or monitor the safety regulations associated with laser-based devices. Class (IV) laser, i.e., 500 mW output is safe, can be prevented from direct contact. Not only laser, but also fumes generated from working cabinets are dangerous. However, processed gases are extracted from cabinets via particle settlement. Resultant clean air is legally allowed to environment.

3.3.5 Laser Engraving

Laser format systems are ideally suitable for cutting of textiles. In the modern era, spacer fabrics, modern fabrics, fleece and needle felt, as well as glass fiber fabric, thermoresistant and other technical textiles are unsurprisingly cut by lasers. In specific, the contact-free processing is a crucial argument, done with the help of laser beam in textile industry. Hence, non-deforming cutting and distortion-free cutting attract more buyers. A live, real-time applications of laser can be everyday recognized for its usage in the textile market [23].

Material Information for the Laser Cutting and Engraving of Textiles

Textiles contain natural and synthetic materials in which laser is used to cut textile likely fabrics, knitted fabrics, meshworks, sew fabrics, nonwoven fabric, and felts. The basic component of them is fibers which are characterized by a high flexibility and suppleness. Textile products like fabric and craft are widely used for various purposes. Among them, medical technology, clothing industry, home textiles, or decorative fabrics are very popular in advertising. Technical textiles are one kind where textiles are used for industrial purpose in which laser plays as filters, upholstery, heat protection fabrics, textile components in the automotive industry or sails.

Laser engraving is a subset of laser marking brightly used to engrave an object. It sounds very complex and computerized to drive the laser head [24]. Despite complexity, very precise and physical contact kind of precision in clean engravings without tearing fabric can be obtained with high rate of production. However, marks produced by laser engraving are permanent. In addition, implementing laser technology shows greater versatility in material selection. This technique seems faster than the non-laser methods used for product imprinting. In some cases, laser is equipped with engraving machine to cut through thin materials and to carve the printing screens, for hollowing, for creating pattern buttons in denim, etc. The embroidered pattern in the fabric or leather or any value-added products can be color faded by burning them in suitable condition.

Gas lasers are customarily used in laser engraving, specifically low-cost CO_2 lasers in denim engraving. Minute designs and patterns are easy to draw, mark, or imprint on finished denim using laser technology. This is more fortunate over sandblasting and acid washing. The design flexibility and accuracy in laser engraving let heavy competition for conventional techniques. Any 3D effects in embroidering or even apparent carves are aided by a computer-aided design (CAD) [25] via suitable laser system.

Work Sequence of Laser Engraving of Denim

1. Creating new patterns or designs by computerized program from captured photograph can be edited using online tool and software for laser engraving.
2. Stored image is converted into gray scale; image format; bmp format (bitmap image format).
3. The file is now processed into the laser system.

4. The parameters for pixel time and resolution are fixed accordingly.
5. Finally, the garments are placed on a honeycomb cutting table.
6. By conducting laser engraving on the identified onto the garments is finally done.

A laser engraving system Fig. 3a, b shows working cabinet of laser engraving system [26].

A honeycomb structure of working cabinet is constructed with tiny holes for fume extraction. Those holes are to circulate gases and to remove debris from the work area. Through suction, fabric can be processed in the working area.

Designing Software

Nowadays, online programs are available to acquire knowledge on different materials for designing and processing like TechShop (e.g., http://www.techshop). Figure 4 deliberately provides you knowledge on what is happening with fabric (Fig. 5).

Since the laser cutters at the TechShop are made by epilog which can be browsed here (http://www.epiloglaser.com) for a free design for fabric. New users can make smaller design just to test for learning.

Steps

a. Place a piece of upholstery fabric underneath the upholstery fabric.
b. Use laser etching as a safeguard with different materials. Laser may be heavy for freshers.
c. Let the machine do its duty.

(a)   **(b)**

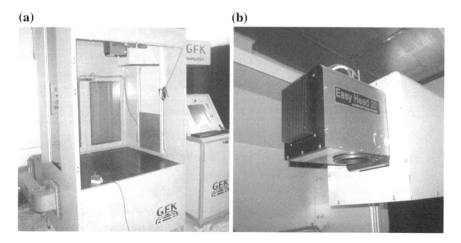

Fig. 3 a, b Laser engraving system with lens

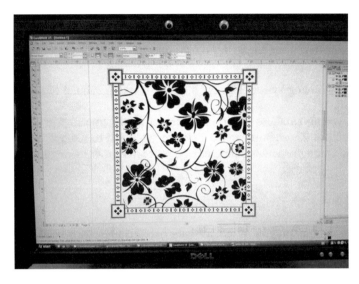

Fig. 4 An image of digital design

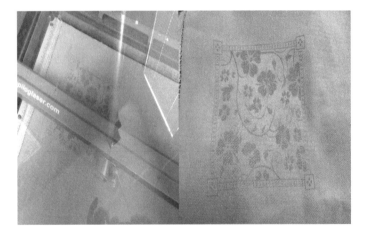

Fig. 5 An image of finished garment

d. Check out the result. There may be slight color change.
e. Remove the plush to reveal harsher weave.
f. Try out more on (http://www.techshop.ws).

While using lasers, color change of fabric is possible while adjusting parameters like power stability, control of laser power, and fast pulse rise time. It may result in reducing garment's life and its sustainability.

Benefits of Laser Engraving and Cutting

- Textile artwork is an art using natural and synthetic fibers to construct practical or decorative objects.
- Positioning on the surface of fabric and crossing over pockets and zip are a kind of ease operation.
- Vertical sleeve or decorating length can be modified at any time.
- Great consistency and precision.
- Laser technology also creates clean-cut edge by applying heat to the laser without any pressure on the fabric. It perceptually avoids the damage of fabric.
- Laser technology is a great challenge for other methods like screen printing, since it is fast in production.
- Laser technology is a relief as contact-less.
- Laser technology assures no bent/fraying in fabrics during laser cutting.
- Simple to use and compatible to incorporate with designing software.
- Integrating laser technology in one's own garment industry is greatly running in no time [27].

3.3.6   Laser in Welding Garment Production

A process of appending fabrics is known as welding for garment production. In case of welding, heat is applied to join the thermoplastic materials together. Though, the sealed garment seems weaker, yet, offers well-finished appearance, flexibility as it does not contain bulky seam [28].

3.3.7   Bar Code Scanning

Identifying each garment product by laser scanners is important to minimize the wastage. For this purpose, barcodes are used with the help of gas lasers. Remarkably, helium–neon (He-Ne) lasers are used to identify unique barcodes by focusing laser beam at each product followed by modulating the signal received from beam and transferring it to computer which contains the product information. Semiconductor-based lasers are widely used for scanning.

In modern days, radio frequency identification (RFID) tags are used rather than bar codes. RFIDs are more advantageous, said by manufacturers from garment industry. Because, it avoids the damage caused by physical handling, quick response, standards of the product as in barcode systems [29].

3.3.8   Laser Marking

Another advantage of laser is to mark products on various surfaces. It contributes fast and clear marking of product with high precision of varying contour and hardness. Significantly, a variety of organic polymers can be processed complex designs with high precession. Laser marking is the best option for first graded clothing and fashion accessories [30].

3.3.9 Miscellaneous Uses of Lasers in Apparel Production

Stitching machine aided by laser beam is used for automatic positioning and creating slit across the fabric. It also identifies, if any needle while stitching is stuck in the final finished garment. For decoration purpose, high-degree laser engraving system is used. A final phase of 'garment finishing' is 'grade fixing' to create a brand new look. Laser technology is upgraded with various surface ornamentations. It does not require wet processing. Moreover, antimicrobial properties enhanced textile seems faultless and popular. Research is extended to stacking nanoparticles on cotton fabric [32].

4 Hazards

Laser has to be labeled in accordance with notable hazard, which may vary based on the class of the laser, the power output, and the wavelength. It has to be legalized in all textile industries. Hazard is stated low when the radiant power output is less than 1 mw. Similarly, lasers with an output more than 1 m are stated high and harm. It is must to have basic knowledge on laser products; how hazard are they based on the risk of injury to the eye?

4.1 Retinal Injury

High concentration of laser on retina can affect the most susceptible body part of eye. Class 4 lasers can cause severe damage likely tissue injuries in the eye interior. Therefore, safety or protection measures have to be taken at processing unit for labors. Because labors are supposed to be affected by 'aversion response' as laser beam shines onto their eyes and collapses 'the blinking reflex.' Therefore, such human instinct may lead to eye injuries. Laser becomes potentially more harm at the time of direct stare at laser. It also results in permanent damage to the eye and subsequent partial or total loss of vision. Consequently, laser beam shining into the eyes forms 'dazzle effect' in which exposure leads to disorientation. However, labeling laser products must meet the output restriction of less than 1 mW [33]. Beyond this limit, aversion response cannot protect eyes from damage. Anyway, various classes of lasers reported within a hazard category may cause eye damage. The class of laser and period of exposure are deciding factors to cause damage with the range of wavelength in the visible spectrum (400–700 nm). Class 3B or Class 4 lasers are responsible for aversion response. Experts suggest that acquiring knowledge on laser and hazard labeling; not to point lasers at eyes; even let laser-based toys away from children may help in recovery.

4.2 Protection

Protections measures must be developed in laser using are by administrative controls. Particularly, Class 3B or Class 4 lasers are more harm; they need personal protective equipment for skin and eye injury in working area.

4.2.1 Eye Protection

Laser control area must be provided with protective eye wears as the maximum permissible exposure (MPE) limit [34] is exceeding if the beam is directly seen by workers. Eyewears include spectacles and face shields, or eyewear using reflective coatings can be taught to workers how to use or wear for the safety measures. Standard operating procedure must be drawn appropriately in favor of workers when unfortunate hazard arises by eyewear instead when the eye protection is not worn. A single type of eyewear does not offer maximum protection from all type of laser radiation and wavelength. Uncertainly, eye protection wear should provide visibility to withstand the maximum wavelength; maximum power of laser radiation; absorption efficiency; optical density with the maximum power rating in combination.

In addition, laser control unit must be inspected periodically to ensure that any damage like splitting and cracking and other damage will not endanger the wearer. A range of laser can be tuned through its wavelengths which causes additional problems. Brand of laser protection goggles must be carefully chosen and handled with great care. For the suitability of a particular type of eye protection as a result of staring class of laser, laser safety officer (LSO) must be available in each production unit.

4.2.2 Accessible Emission Limit (AEL)

Accessible emission limit (AEL) of various classes of laser relates to health hazard that is discussed in this section. Power and energy emitted by laser are accessible by user. Exceeding AEL and MPE limits by eye or skin either directly from the beam or after reflection is more dangerous [35]. The route of laser emission sources is range of wavelengths, energies, and pulse durations. Characteristics of laser can bring better understanding of health hazards causing by its classes.

Class 1 laser does not exceed MPE values due to its inferior intensities. Class 1M laser is carrying low intensities with high power, causing a danger hazard. Class 2 in the visible range is harmless for eye. Because bright nature of this laser makes user to close or blink eyes just to escape the bright light. Class 2M sources carry low intensities with high power, causing serious danger. Medium power is carried by class 3 laser, causing a short skin exposition. Direct eye exposure to class 3R laser is five times higher than class 1 sources. Class 3B sources do pass medium power, causing severe damage if the exposure exceeds 10 s. The direct vision of class 4 laser beam is dangerous to eye [34, 35].

4.3 Skin Injuries

Skin injuries from lasers are less severe than the eye injuries. But the rate or chance of exposure to laser radiation is high due to its large surface area. However, losing vision is irreparable compared to skin injury. Moreover, exposing skin to laser radiation is easily preventable. Hereby, thermal injury causes burn from acute exposure of high-power laser beams. Photochemically induced injury causes chronic exposure to scattered ultraviolet laser radiation. In some cases, specular reflections can even lead to photochemical injury over time. Yet they are painful, regardless not severe, and easy to prevent hazard awareness. Through the prolonged exposure to the direct beam, specular reflections cause severe sunburn and more consequently to the formation of skin cancer.

4.3.1 Protection

The safety measures on hazard control stop entering laser to skin or eye. Appropriate eyewear, clothes, and covers for forearms are often necessary to control hazard. Administrative control inspects standard operating procedures (SOPs), engineering controls, and protective controls periodically. Controlling measures of UV exposure to eye or skin demand LSO to provide specific information or guidelines. Laser escaping from the process is being prevented by window drapes in the laser control unit. Choosing appropriate window drape needs basic understanding of specific class of laser. Sometimes, window drapes are not required if gas baser laser cannot penetrate glass.

4.4 Electrical Shock

Electrocution or electrical shock is mainly threatening workers who work under high voltages of power supply in laser control unit. Laser electronics is life threatening as compared to exposure of laser beam. Lethal voltages are baleful when one is not experienced working with high voltages. A well-versed personnel can be allowed for working at high laser power supply. During maintenance, the power supply should be unplugged from its electrical outlet [36].

5 Conclusion

Everyone's wardrobe is globally filled with trendy and timeless jeans. But people having no idea about the environmental consequences of these are real wonder. As garment industry requires huge amount of water, energy and chemical for dyeing can induce environmental impacts. Researchers expect in near future that sustainable development in value-added products is necessary to implement. Denim in exemption attracts more customers for its type of shades and finishes with harmless productive techniques. Sustainable denim finishing with 3D effect is advantageous over stone/mill wash, moon wash, and bleach. Customers' prospective in denim market is whisker effect by dry finishing.

On the other hand, purchasing an 'n' number of pair of jeans is equivalent to wasting tons of water. It also forms pavement for greenhouse gas emissions. Economical, ecological, denim products must be introduced by developing sustainable garment industries. Small- and medium-scale industries have to be supported by government/non-government financial schemes to promote laser technology rather than mechanical abrasion, ozone fading, waterjet fading, tinting technique, etc. But all of all are environment-friendly using less water and energy. Aforementioned laser-based technologies are to facilitate sustainable innovation in the field of textile design and manufacture, enabling transition toward a circular economy. In combination, they can be implemented for the sustainability of products.

References

1. Nayak R, Khandual A (2010) Application of laser in apparel industry. Colourage 57:85–90
2. Dowden J (2009) The theory of laser materials processing: heat and mass transfer in modern technology. Springer Science & Business Media, Berlin
3. Nayak R, Singh A, Padhye R, Wang L (2015) RFID in textile and clothing manufacturing: technology and challenges. FATE 2:1–16
4. Kan C (2014) Colour fading effect of indigo-dyed cotton denim fabric by CO_2 laser. Fiber Polym 15:426–429
5. Lucas J, Belino N, Miguel R et al (2015) Digital printing techniques for denim jeans. Denim: Manufacture, Finishing and Applications, p 287
6. Martínez-Sala AS, Sánchez-Aartnoutse JC, Egea-López (2013) Garment counting in a textile warehouse by means of a laser imaging system. Sensors 13:5630–5648
7. Sarkar J, Rashaduzzaman M (2014) Laser fading technology: facts and opportunities. Textile Today
8. Final activity report—ALTEX/R/TWI/IAJ/07122021
9. https://www.troteclaser.com/en/applications/textiles/
10. Mathews J (2011) Textiles in three dimensions: an investigation into processes employing laser technology to form design-led three dimensional textiles. Thesis submitted to Loughborough University, pp 54
11. EC CONTRACT COOP-CT-2005-017614
12. https://www.engadget.com/2014/06/17/laser-cut-clothing-explainer/

13. Mallik-Goswami B, Datta AK (2000) Detecting defects in fabric with laser-based morphological image processing. Text Res J 70:758–762
14. http://www.fabriclink.Com/university/burntest.cfm
15. http://www.wzl.rwth-aachen.de
16. Islam A, Akhter S, Mursalin TE et al (2006) Automated textile defect recognition system using computer vision and artificial neural networks, WASET 13 (ISSN 1307-6884)
17. https://www.trotec-materials.com/laser-materials/foil-sticker/laserflex.html
18. Solaiman, Saha J (2015) Comparative analysis of manual fading and laser fading process on denim fabric. SD 3(6):44–49
19. Mathews J (2011) Textiles in three dimensions: an investigation into processes employing laser technology to form design-led three dimensional textiles. Thesis submitted to Loughborough University, pp 84–96
20. Water free technology for denims at Fibre2Fashion.com
21. Yuan G, Jiang S, Newton E et al (2012) Fashion design using laser engraving technology. In: 8ISS symposium-panel on transformation pp 65–69
22. Choudhury I, Shirley S (2010) Laser cutting of polymeric materials: an experimental investigation. Opt Laser Technol 42:503–508
23. Hung O, Song L, Chan C et al (2011) Using artificial neural network to predict colour properties of laser treated 100% cotton fabric. Fiber Polym 12:1069–1076
24. Petrie E (2015) Alternative fabric-joining technologies 13. In: Nayak R, Padhye R (eds) Garment Manufacturing Technology Cambridge. Elsevier, Cambridge
25. Laser joining fabrics improves productivity published on 01/01/2005
26. http://www.techshop.ws
27. Chapter 6: Innovating clean energy technologies in advanced manufacturing. Quadrennial Technology Review 2015
28. Nayak R, Padhye R (2014) Introduction: the apparel industry. In: Nayak R, Padhye R (eds) Garment manufacturing technology. Elsevier, Amsterdam
29. Fan J, Liu F (2000) Objective evaluation of garment seams using 3D laser scanning technology. Text Res J 70:1025–1030
30. https://www.apparelviews.com/growing-application-laser-apparel-industry/
31. www.ARPANS.com
32. Section III: Chapter 6, Laser hazards. Published by Unites States, Department of Labour Safety
33. Personal Protective Equipment published on Environmental Health & Safety
34. http://www.optique-ingenieur.org/en/courses/OPI_ang_M01_C02/co/Contenu_08.html
35. Sliney DH (1995) Laser safety. Lasers Surg Med 16:215–225
36. Wyrsch S, Baenninger PB, Schmid MK (2010) Retinal injuries from a handheld laser pointer. N Engl J Med 363:1089–1091

Sustainable Wastewater Treatment Methods for Textile Industry

Aravin Prince Periyasamy, Sunil Kumar Ramamoorthy,
Samson Rwawiire and Yan Zhao

Abstract All over the world, environmental considerations are now becoming vital factors during the selection of consumer goods which include textiles. According to the World Bank, 20% of water pollution globally is caused by textile processing, which means that these industries produce vast amounts of wastewater. Generally, these effluents contain high levels of suspended solids (SS), phosphates, dyes, salts, organo-pesticides, non-biodegradable organics, and heavy metals. Increase in water scarcity and environmental regulations has led to textile industries to seek for sustainable wastewater treatment methods which help to reduce their water footprint as well as reduce their operational costs. Therefore, sustainable wastewater treatment could be the best choice for the textile industries with respect to the current issues. So, it is important to discuss and champion awareness mechanisms which help to reduce the current issues with respect to the textile wastewater. Therefore, this chapter intends to discuss the various sustainable wastewater treatments, namely granular activated carbon (GAC), electrocoagulation (EC), ultrasonic treatment, an advanced oxidation process (AOP), ozonation, membrane biological reactor (MBR), and sequencing batch reactor (SBR).

Keywords Wastewater · Effluent · Textile industry · Electrocoagulation
Water pollution · Pollution · Membrane bioreactor · Ultrafiltration
Activated carbon and sustainability

A. P. Periyasamy (✉)
Department of Materials Engineering, Technical University of Liberec, Liberec,
Czech Republic
e-mail: aravinprince@gmail.com

S. K. Ramamoorthy
Department of Mechanical Engineering, University of Boras, Boras, Sweden

S. Rwawiire
Department of Textile and Ginning Engineering, Busitema University, Tororo, Uganda

Y. Zhao
College of Textile and Clothing Engineering, Soochow University, Suzhou,
People's Republic of China

© Springer Nature Singapore Pte Ltd. 2018
S. S. Muthu (ed.), *Sustainable Innovations in Apparel Production*,
Textile Science and Clothing Technology,
https://doi.org/10.1007/978-981-10-8591-8_2

1 Introduction

Among all planets in the solar system, the earth is notably a unique planet because of the life-supporting conditions present. Life forms of various nature, from microorganisms to human beings, dwell on earth. The environment that supports life and sustains various human activities is generally known as the biosphere. The term "environment" means surroundings. It is defined as "the sum of all social, economic, biological, physical, or chemical factors which constitute the surroundings of man whose task is to govern the environment." The biosphere is a shallow layer compared to the total size of earth and extends to about 20 km from the bottom of the ocean, the highest point in the atmosphere at which life can survive, without man-made protective devices. Man's habitation of the earth and his activities such as agricultural and industrial revolutions followed by the current high-tech world of synthetic and man-made materials have had a negative impact on the environment and biodiversity. Some the major impacts of man on the environment are discussed briefly below. Reduced child mortalities and developments in medicines had increased the lifespan of humans. The birthrate/m is 2–3 times higher than the death rates; with these rates, by 2050, the world's population is expected to be around 9.8 billion compared to the current level of 7.6 billion [1]. This population stress leads to increase land habitation and demands, on resources like water, minerals, fossil fuels, and food, etc., all at the cost of major environmental degradation. Ongoing depletion of ground and surface waters, natural wealth, fossil fuels, forests, greenery and species diversity, due to overconsumption habits of human, is a permanent threat posed by humanity and the environment. There is rapid increase in deforestation which eventually leads to soil erosion, loss of plants and niche to other species, global warming and climatic changes. Deforestation, hunting, urbanization, and agricultural land reformation lead to the extinction and deprive the habitat to a vast variety of species, leading to endangered species and animals. However each and every biotic element has a role in nature's harmony, the ongoing loss of biodiversity is a threat to natural ecosystems. The massive urbanization brings in the need for housing, utilities, buildings, industries, roads, transportation, infrastructures, water, and energy supplies. All this is attained at the cost of deforestation and loss of greenery. The end results are congestion, water, energy demands, pollution issues, and loss of biodiversity. Overconsumption of water for domestic, agricultural, and industrial needs ends in contamination of water sources, surface and groundwater depletion, and degradation. Water, a major abiotic element for support of life, is rendered useless for consumption by way of organic, mineral, toxic, bacterial, and physical pollutions. Water pollution is an alteration of physical, chemical, and biological properties of water in a water body, rendering it unsuitable for consumption. Water pollution renders it unsafe for human or animal consumption, for industry, aquaculture and agriculture or recreation leading to land, thermal, and marine pollution. The need for energy for domestic, transport, and industrial operations has, for instance, led to combustion of fossil fuels in due process delivering pollutant greenhouse gases like carbon monoxide, carbon dioxide, nitrogen, sulfur, and carbon particles and add to the air temperature. Industrial emissions and chlorinated products also add potential contaminants into the atmosphere, the end results being acid rains, global warming, air toxicity, climatic change, and depletion of ozone layer. Deforestation and urbanization are the major causes of land abuse and soil erosion in due course leading to loss of fertility, contamination of pollutants—solid and liquid wastes on to land. The industrial revolution led to the birth of various industries, among them the textile industry. Textile industries have two divisions:

- Dry process—mostly engineering and assembling departments, which do not use water for all practical purpose (e.g., blowroom, carding, weaving).
- Wet process—departments which use water as raw material or in their processes or for both (e.g., chemical processing like coloration, printing, garment washing).

A couple of countries has the largest share of their gross domestic product (GDP) coming from textile and apparel industries; however, a hasty and unplanned clustered growth of these industries in most cases has environmental concerns. This chapter aims to present environmental challenges in textile industries and importance of sustainable wastewater treatment. Furthermore, it includes the comprehensive discussion of various conventional effluent treatments as well as its pros and cons as compared to various sustainable wastewater treatments, namely granular activated carbon (GAC), electrocoagulation (EC), ultrasonic treatment, advanced oxidation process (AOP) such as ozonation, membrane biological reactor (MBR), sequencing batch reactor (SBR), and rotating biological contactor (RBC). This chapter merely attempts to discuss the various sustainable wastewater treatments in textile industry, namely GAC, EC, ultrasonic treatment, AOP, ozonation, MBR, and SBR. This treatment could be a best choice for the textile industries with respect to the current environmental issues. Generally, the conventional wastewater treatments not only increase the landfill and extensive infrastructure needs, but also it changes the ecosystem when the landfill is not properly managed. In addition, these technologies consume considerable energy, which consequently emit the carbon emission. However, sustainable wastewater treatments provide significantly increased pollutant removal efficiency, and considerably reduce the sludge management, energy consumption, and running cost. In this process sequence, biologically based treatments offer high organic removal efficiency and adverse impact on the ecosystem, and also it ensures the sustainable sludge management which consume less energy; therefore, it reduces the overall cost.

2 Water Pollution Associated with Textile Industries

The coloration of textile involves a group of dyes, namely acid, direct, reactive, metal: complex, vat, sulfur, disperse and pigment. Generally, colorants are complex of organic or inorganic chemicals, and it can be applied to textile with different

Fig. 1 Textile effluent contaminates of Noyyal River at Tirupur, India

methods; however, the dye exhaustion is varied from 50 to 85%; therefore, dyes will discharge in the effluent as a pollutant, which badly affects the nearby water streams (Fig. 1). The color of dye wastewater changes the color of entire pollutant; however, it looks brownish red, black, or a mixture of both, and it also contains 4000–5000 parts per million (ppm) of SS [2–5]. Apart from a colorant, huge quantity of chemicals and auxiliaries are used during wet processing; in addition to process water, large volumes of cooling water are discharged as waste [6–10]. Apart from these effluents, solid wastes and liquid wastes as effluent are also discharged on land or into sewers or into natural water resources. The various effluents which are generated from textile chemical processing are given in Table 1, and the domestic standards for effluent discharge are given in Table 2. The effluent discharge standards for Indian textile industry are given in Table 3.

2.1 Effluent Discharged Various Operation of Textile Process

Melt and dry spinning operations produce not much effluent as compared to wet spinning [13]; however, wet spinning, for example, of viscose rayon produces huge pollutant in the wastewaters [14]. Table 4 shows the possible pollutant which caused by wet spinning and dry spinning. The basic steps involved in viscose rayon manufacture are wood pulping, soda cellulose formation and molecular fragmentation, solubilization with carbon disulfide, blending, wet spinning, washing, and finishing [15].

Table 1 Various pollutions caused by textile industry

Acids	pH reduction causes corrosion, help in the liberation of Na_2S, increase TDS, destroy microorganism, destroys self-purification system of water
Alkalis	Increase the pH, helps in the liberation of hydrogen ions, increase TDS, destroy microorganisms, destroy self-purification of water
Dissolved solids	Chlorides, sulfates, nitrates, bicarbonates of various inorganic metals, plants could be affected by osmotic changes, affect biological organisms, increase the COD and BOD
Suspended matter	Insoluble wastes, destroys self-purification of the stream of water, reduces the photosynthetic activity of plants, chokes the gill of fish, leads to floating masses, offensive odor
Dyes	Change the color of water, increase the COD and BOD, lead to cancer, disabilities, and disorders
Oils	Absorption of O_2 from air into water affected, destroys the animal, plant life in water, increases the COD and BOD
Synthetic detergents	Reduce biodegradability, toxicity, accelerate chain reactions, increases floating masses, and affect natural aeration
Toxic metals	Self-purification of streams destroyed, cause skin diseases, disabilities, and disorders, destroy microorganism aquatic plants and animals are killed, compounds containing mercury affect the food chain
Pesticides	Fatal to fish, metabolic activities of body and photosynthesis of plants are affected
Gaseous pollutants	Include ammonia, free chlorine, hydrogen sulfide Increase toxicity, add on to oxygen demand, kill microorganism, destroy animals and plant kingdom
Heated effluents	Reduce dissolved oxygen concentration, aquatic life gets affected, destroy animal and plant kingdom, affect the entire ecosystem because of temperature
Radioactive wastes	Ill effects are numerous, induce metabolic changes, disorders, disabilities, genetic damage, chronic diseases, cancer, the abnormal birth of human beings which are also common to plants and animals

Table 2 Domestic effluent discharge standards [11]

Parameter	Inland surface water	Public sewers	Inland for irrigation	Marine coastal areas
pH	5.5–9.0	5.5–9.0	5.5–9.0	5.5–9.0
TSS (mg/l)	100.0	600.0	200.0	100.0
TDS (mg/l)	2100	–	2100	–
BOD (mg/l)	30.0	350.0	100.0	100.0
COD (mg/l)	250.0	–	–	250.0
Sulfates (mg/l)	1000.0	1000.0	1000.0	–
Chlorides (mg/l)	1000.0	1000.0	600.0	–
Oils and grease mg/l	10.0	20.0	10.0	20.0
Lead (pb) (mg/l)	0.1	1.0	–	2.0

Table 3 Limits to discharge for textile chemical processing in India (adapted from [12])

Parameters	Concentration not to exceed, milligram per liter (mg/L), except pH
Chemical oxygen demand (COD)	250
Biological oxygen demand (BOD)	30
pH	6–9
Total suspended solids (TSS)	100
NH_3	15
Sulfide (S)	1
Total residual chlorine	1
Total chromium as Cr	2
Aniline	1
Oil and grease	10
Phenolic compounds as C_6H_5OH	1

Table 4 Effluent discharged by man-made fiber industries

Origin	Characteristics
Wood pulping	Alkali + surfactant (high TDS, BOD, COD)
Xanthation, dissolving	Carbon disulfide, alkali solution (high COD, alkalinity)
Spinning	Carbon disulfide, sulfuric acid zinc sulfate, sodium sulfate (high acidity, TDS, turbidity), solvents used in dry spinning
Washing and finishing	Sodium sulfide, sodium hypochlorite, hemicellulose (TDS, high BOD, COD)

As discussed earlier, chemical/wet processing of textile industry alone contributes 70% of the pollution. It is well known that textile wet processing consumes large amount of water for a various processes like sizing, desizing, scouring, bleaching, dyeing, printing, finishing, and other treatments (Fig. 2); further still, the different chemicals utilized lead to an acidic pH (Tables 5 and 6), and eventually, a large quantity of wastewater is discharged as a pollutant, which enormously spoils the water streams (Table 7).

Manufacturing of dyes and pigments deals with aromatic hydrocarbons amines, mineral acids like sulfuric and nitric acids and also involves a chain of reactions like starting toxic raw materials, intermediate steps like diazotization, coupling, sulfonation, nitration, etc.

The effluent from the dye industry is of high organic content, colored, and of high COD/BOD values. The dye and pigment industrial wastewaters are composed of insoluble organic pigments, dyes, heavy metals, emulsifiers which are shown in Table 8.

Fig. 2 Percentage share of water consumption in chemical processing of cotton materials

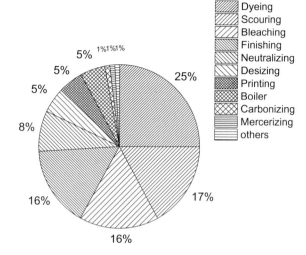

Table 5 General process parameters to be determined in pre-treatments

Parameters	Sizing	Desizing	Scouring	Mercerizing	Chlorite bleaching	H_2O_2 bleaching
pH	7–9.5	7.3	10–13	14	6	10.5
TDS (mg/l)	–	8700–10,200	6323	12,000–5000	22,000	2400
SS (mg/l)	240–260	200–270	55	2200	6500	420
BOD (mg/l)	600–2500	4400–5060	2100	45–65	100	75
Color	Cream	Colorless	Brown	White	White	White
Odor	Bad odor	Bad odor	–	–	Bad odor	–

Table 6 General process parameters to be determined at coloration and combined effluent

Parameters	Dyeing	Printing	Combined effluent
pH	5–12.5	4–9	8–9.5
TDS (mg/l)	35,000	2000	20,000
Sus. solids (mg/l)	26,000	15,000–20,000	10,000
Total solids	40,000	2500	30,000
BOD/COD (mg/l)	300/1500	400/2000	250/400
Temperature °C	40–50	50	35
Heavy metals (mg/l)	450	400	420
Total alkalinity (ppm)	1600	–	–
Oils and greases (mg/l)	1900	–	3.6
Color	Heavily colored	Heavily colored	Heavily colored

Table 7 Effluent discharged by textile chemical processing

Origin	Characteristics
Sizing	Starch, colored, high BOD, suspended solids, PVA, odor, softener, oils fats [16]
Desizing	Starch, CMC, PVA, hydrolyzed starch, enzymes, salt, acidic pH [16–18]
Scouring	Alkalis, surfactants, saponified oils, hydrolyzed pectins, proteins, suspended solids, oil, high pH, silicates with high COD, natural colors, and TDS
Mercerizing	High alkali, suspended solids, TDS
Bleaching	Chlorines, hypochlorites, alkali, peroxides, silicates, suspended solids, fatty alcohol [16]
Dyeing	Dyes, salt, alkalis, acids, detergents, chromium, copper, high BOD and COD, and TDS [20–23]
Printing	Dyes, alkali, acids, chromium, copper, thickeners polymers, detergents high BOD and COD, waxes, oils fatty alcohol
Finishing	Silicons, suspended and dissolved solids, cationic compounds organic and inorganic compounds
Carbonizing	Carbonized cellulose, high acidity, TDS
Synthetics	Hot color, high BOD & COD, alkali, organic solvents, acidity, TDS, peroxides, hydrolyzed PVA, acrylics, etc.
Weight reduction	High alkalinity, turbidity, high BOD/COD
Wool scouring	Hot, highly colored, high BOD and COD, suspended and dissolved solids, grease, soaps, and alkalis
Silk degumming	Dissolved solids, high BOD, high turbidity, bad odor, TDS

Table 8 Effluent discharged by dyed and pigment manufacturing industries

Origin	Characteristics
Dyes	
Diazotization coupling reaction, nitration, sulfonation, halogenations, oxidation, reduction, and condensation reactions	Highly colored, high organic content, high BOD (200–400 mg/L), high COD (500–2000 mg/L), high BOD/COD ratio (0.2–0.4), acidic pH, high TDS, the presence of heavy metals like chromium, copper, cobalt, etc.
Pigment	
Grinding ball mills, washings	Highly colored, BOD, COD, turbidity of organic character [22]

3 Effluent Treatment Process

Wastewater from the textile industry is the main source of the organic contamination, which causes hemorrhage, ulceration of skin, nausea, skin irritation, and dermatitis. The contamination present in the effluent can block the sunlight and increase the BOD, therefore preventing photosynthesis and reoxygenation process. Various processes need to be implemented so as to reduce the pollution load; in this case, many physical methods (adsorption, filtration methods, coagulation, and flocculation processes), chemical methods (oxidation, advanced oxidation, Fenton's reagent), and biological treatment (anaerobic, aerobic) are carried out. The whole effluent process involves three steps (Fig. 3), namely primary treatment, secondary treatment, and tertiary treatment. In general, the primary treatments involve removal of the SS, floating and gritty materials, the secondary treatment involves reduction

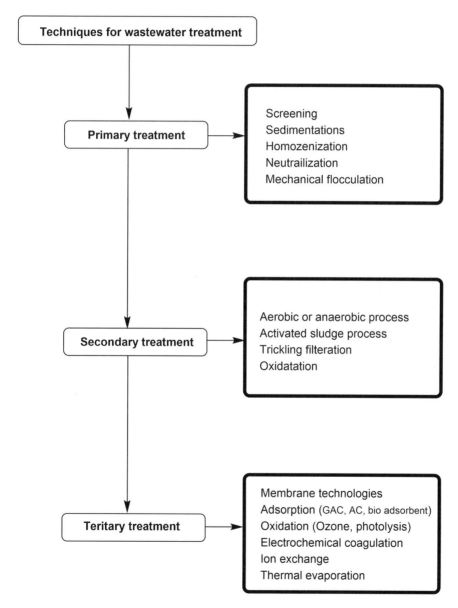

Fig. 3 Wastewater techniques for textile wet processing industries

of oxygen demands, other chemicals, and color of pollutant, and lastly, the tertiary treatment involves the removal of the final contaminates left over in the pollutant. After the treatments, treated water must be analyzed for the quality so as to gage its reuse or discharge into the water streams [23–25].

Sustainable Wastewater Treatment and Overview of Existing Color Removal Methods

As discussed earlier, water and land is vital but is a limited resource, and it should not be polluted as a result of the industrial process. Therefore, the main task for the emerging technologies is to protect the earth from the pollution of industrial processes. In the present scenario, the conventional effluent treatment discharges huge quantities of sludge, which is an unwanted residue; perhaps sludge management is critical with respect to the environmental issues [23, 24, 26–28]. Sludge eventually ends up in landfills; however, it is not a safe practice since it leads to the pollution of air, water, and soil, also generates the uncontrolled landfill gases like methane, CO_2, and volatile organic carbons were proceed to global warming. During the landfilling, there is a huge possibility to contaminate the local groundwater due to the leaching potential [29]. Groundwater contamination from sludge is also an environmental concern and, therefore, quest for an alternative technology that can reduce the sludge during the effluent treatment is a step in the positive direction. In this case, chemical coagulation and other conventional effluent treatment process contribute significant impact on the sludge formation; therefore, it is advised to replace by electrocoagulation (EC) process. The colored water from textile effluent could be the biggest threat to the environment, and there are 12 classes of chromogenic groups, but azo and anthraquinone types are mostly used in the textile industry; both classes contribute 80% of total dye contributing to the textile colorants. On average, the dye house effluent contains 0.6–0.8 g dye 1^{-1} [30], hence the color and dye removal in the effluent becomes more important and it is a major scientific interest. In order to remove the colors from textile wastewater, various methods are used, such as chemical oxidation, coagulation and flocculation, membrane separation, aerobic and anaerobic degradation.

As there are some limitations such as the excess amount of chemical usage, accumulation of concentrated sludge that has serious disposal problems and lack of effective color reduction in some of these techniques, it has been shown to be effective. Based on the transfer of pollutants from the solution to the solid phase, it is known as the adsorption technique which is one of the efficient and general wastewater treatment methods. In terms of the initial cost, the simplicity of design, ease of operation, and non-toxicity of the utilized adsorbents, the adsorption method is superior to other dye removal techniques such as conventional wastewater treatment methods [31]. In selecting the absorbent to remove organic compounds from wastewaters [31, 32], the main criteria are the cost-effectiveness, availability, and adsorptive properties and especially non-toxic and green adsorbent-based application of adsorption procedure with high surface area, and the reactive surface atom is a great demand [33, 34]. The existing process and alternative process are given in Table 9.

Table 9 Existing and alternative methods for textile effluent treatments

Existing methods	Alternative methods
Emulsion breaking	Multimedia filtration
Gravity-assisted separation	Carbon adsorption
Skimming	Ion-exchange
Plate/tube separation	Air-stripping
Dissolved air flotation	Activated sludge
Chromium reduction	Biological film or trickling filters
Cyanide destruction	Membrane bioreactors
Chemical precipitation	Electrodialysis (EC)
Filtration	Ultrafiltration
Sand filtration	Reverse osmosis

3.1 Primary Treatments

The purpose of primary treatment is to remove insoluble, invisible, and colloidal organic color and SS, and this treatment involves screening, equalization, neutralization, and coagulation followed by sedimentation steps. Primary reductions of COD and BOD value are also enhanced by this treatment. Various steps of primary treatment are as follows:

Screening
This process can remove the floating and suspended materials, and screens of various sizes and shapes are used depending on the nature of solids to be removed. Generally, fixed bar screens are the most used. Other simple screening equipment is mix mesh frames, alternate perforated plate frames, or damper plate with specific spacing. The screenings are disposed of by various methods such as grid, incineration, grinding, and digestion.

Equalization
The textile effluent has the difference in characteristics like color, turbidity, pH, BOD, COD, TDS, etc. Prior to the primary treatment, it is necessary to proportionately mix various process water to get together to achieve a uniform character. Equalization helps in providing such uniform effluent and prevents shock loads and sudden increase or decrease in pH due to some textile processes like mercerizing, scouring, carbonizing effluents entering the plant suddenly. Also, this process improves the dissolved oxygen (DO) levels.

Neutralization
Most of the secondary biological treatments are effective only if the pH value of effluent is between 6 and 9. However, extreme pH conditions make ineffective secondary treatments, since digestive bacteria can survive close to neutral pH conditions and are dead at extreme pH. Therefore, it is necessary to adjust the pH in the range of 6–9. Depending on the situations, dosing at regular uniform rate should

be done from the stored tanks containing the above liquors. Sometimes it becomes necessary to add dilute H_2SO_4 or CO_2 as per need.

Coagulation sedimentation
Coagulation sedimentation is one of the most used methods, especially in the conventional effluent treatment process. However, the action on the suspended matter, the colloidal type of very small size, and their electrical charge give repulsion and prevent their aggregation. In this case, add the electrolytic products (aluminum sulfates, ferric sulfates, ferric chloride) which can eliminate the surface electrical charge of the colloids. As per the environmental concerns, electrocoagulation can replace this process since it is not environmentally friendly.

3.2   Secondary Treatment Process

Presence of soluble organics, toxic metals, and organics may increase the BOD/COD values of effluent. The aim of secondary treatment is digestion of soluble organics by bacteria (aerobic or anaerobic mode), which improves the DO content, therefore resulting in reduced BOD/COD levels (80–85%) and toxicity. So, this process is enhanced by proper pH control (close to neutral), feeding of nutrients, prolonged interaction with air/oxygen, etc.

Aerobic
Aerobic biological treatment (oxidation) is carried out in the presence of air (open condition), so the soluble organics get converted into CO_2 and it could be removed out of effluent. Anaerobic biological treatment (treatment) is carried out under closed conditions in the absence of air; here, organic materials get converted into CH_4 or ammonia and are released out of effluent.

Activated sludge process
In this method, effluents were continuously exposed and subjected to biological degradation carried out by "microbial floc" suspended in the reaction tank, where the oxygen is introduced mechanical manner. The flocks formed in activated sludge process (ASP) are "zoology masses of living organism" embedded with food and slime material and are active centers of biological oxidation and therefore the name activated sludge process. A number of bacteria to be added depend on the organic matter present in the effluent to be treated.

3.3   Tertiary Treatment

The purpose of tertiary treatment is to remove dissolved solids (minerals, salts) from wastewater. The wastewater after efficient tertiary treatment can be safely discharged or reused for processing. The modern effluent treatment plants employ the following tertiary treatment operations to separate dissolved solids and recover pure/safe water for reuse.

3.3.1 Membrane Technology

The membrane is a thin section of a material which arranged in suitable configuration (sheet or roll form); generally, it acts as a phase separator which permeable to one or like components [35]. Membranes are usually made up of polymers like as cellulose acetate/cellulose, triacetate polyamide, polyester, polypropylene, etc. The pore size of the membrane is usually expressed in micrometer/nanometers [36]. There are different membrane systems that could be used in textile wastewater treatment; they are micro-, ultra-, nano-, and RO filtrations, and each one has different cleaning properties (Table 10).

Microfiltration

Microfiltration process is based on the membrane technology; it can remove the particles in the range of 0.1–0.2 μm. The principle of microfiltration is a physical separation. Substances that are larger than the pores in the membranes are fully removed, meanwhile substances with smaller than the pores of the membrane are partially removed [38, 39]. However, SS, bacteria, pigments, oils, and large particles are removed. Microfiltration rejects ranges up to 10%. Microfiltration membranes are made of several polymers including poly (ether sulfone). Poly(vinylidene fluoride) and poly (sulfone). Ceramic, carbon, and sintered metal membranes have been employed where extreme chemical resistance or high-temperature operation can be carried out [21, 40].

Ultrafiltration

Ultrafiltration utilizes membranes to remove usually in the range of 500–1000 molecular weight. Ultrafiltration is generally preceded by conventional filters or cartridge filters to prevent blinding of the membranes. Periodically, the membranes must be cleaned by backflush technique. Ultrafiltration is used to separate polymers from salts and low molecular weight materials, with pores of 0.001 m–0.1 μm.

Table 10 Filtration spectrum of different membranes [37]

Process	Pore size (μm)	Molecular weight	Could be remove
Microfiltration	0.007–2.00	>100,000	Bacteria, pigments, oil, etc.
Ultrafiltration	0.002–0.10	1000–200,000	Colloids, virus, protein, etc.
Nanofiltration	0.001–0.07	180–15,000	Dyes, pesticides, divalent ions, etc.
Reverse osmosis	<0.001	<200	Salts and ions

Turbidity is sharply reduced by 99%. Polymers are retained for reduced Total Oxygen Demand (TOD), BOD, and COD. It is an excellent filtration technique to remove metal hydroxides and heavy metal up to 1 ppm or even less [21, 36, 39].

Nanofiltration

Nanofiltration membrane works similar to reverse osmosis (RO) except for pressure (less pressure is required for nanofiltration, i.e., 70 and 140 psi); it is due to the larger membrane pore size (0.05–0.005 μm). Nanofiltration removes the dyes, pesticides, and divalent ions, and also it is capable of removing hardness elements such as calcium or magnesium together with bacteria, viruses, and color. However, it operated with lower pressure than RO; therefore, operating cost is less than RO process [21, 36, 38–40].

Reverse Osmosis

In RO process, the good water is separated from pollutant through a membrane with the help of osmotic pressure. However, the RO process may be reversed by applying a pressure on the brine side higher than the osmotic pressure. A series of tubes made up of porous material is lined on the inside with an extremely thin film of cellulose acetate semipermeable membrane. These tubes are arranged in a parallel array; brackish water is pumped continuously with high pressure (>25 atm) through these tubes, and due to the high pressure, only water can penetrate through membrane and salt and another pollutant cannot, since the size of the pore can allow only water. Concentrated brine (reject) and freshwater (permeate) are withdrawn through their respective outlets. Prior to RO filtration, wastewater should be treated with sand filtration, activated carbon (AC) filtration, microfiltration, ultrafiltration, and removals of iron/calcium are necessary for smooth functioning. Generally, RO rejects ranges from 40 to 50% which is further run into the second stage and third stage of RO to recover maximum freshwater [41, 42].

Activated carbon treatment
AC is composed of a microporous, homogenous structure with high surface area (one gram of it has a surface area of approximately 500 m^2) and shows radiation stability which is widely used adsorbent in industrial processes. In the developing countries, the process for producing high-efficiency AC is not completely examined; moreover, there are several problems with the regeneration of used AC. Nowadays, there is a large curiosity in order to find inexpensive and effective alternatives to the existing commercial AC. When an effective and low-cost AC is explored effectively, it may contribute to environmental sustainability and offer benefits for future commercial applications. Comparatively, the costs of AC prepared from biomaterials are very low to that of the cost of commercial AC [43–46]. There are numerous studies which can be carried out to produce AC from waste

materials, such as waste wood [47], bagasse [48], coir pith [49], coffee husk [50], pinecone [51], coconut shell [52], and coconut flowers [53]. AC is ordinarily used to adsorb natural organic compounds, taste and odor compounds, and synthetic organic chemicals in drinking water treatment. Both the physical and chemical process of accumulating a substance at the interface between liquid and solids phases is the adsorption. Due to the highly porous material and in providing a large surface area to which contaminants may adsorb, AC is an effective adsorbent. The two important types of AC utilized in water treatment applications are granular activated carbon (GAC) and powdered activated carbon (PAC). AC has the immense surface area and highly porous nature which allows the adsorbed compounds to penetrate the material fully, endeavoring out all available binding points. Chemical reactions also take place to change some offensive compounds into less objectionable variations. Even though the AC is not effective against all compounds, it does have the ability to bond with compounds in all three phases: liquid, solid, and gas.

Production

A wide variety of carbon-rich precursor materials such as bituminous coal, anthracite, sub-bituminous coal, lignite, wood, coconut shells, and peat are used to produce ACs. By either thermal or chemical activation processes, these materials are converted into AC. Steam gasification (activation) and chemical activation used reactive, inorganic additives at relatively lower temperatures and are typically included in thermal treatment. Bituminous coal can be classified as direct-activated or re-agglomerated to produce the ACs. Direct activation is involved to size the coal approximately to the required particle size and thermally activating the sized coal. Comparatively, direct activation can produce a less costly product to that of re-agglomeration. Re-agglomeration involves first pulverizing and briquetting the coal with organic binders. The desired particle size is achieved through the stage crushing the briquettes. The binder is also converted to a graphitic structure that interconnects the activated coal particles when the agglomerated material has been activated. Both the hardness and abrasion characteristics of re-agglomerated and direct-activated GAC are frequently comparable. In order to have a better certain organic contaminant removal, re-agglomerated carbons tend to have a more homogenous pore structure. All the carbonaceous raw material reacts during thermal treatment by means of condensation reactions to form increasingly larger aromatic plate structures. The characteristics of the raw materials influence the density of the structure. In order to have more extensive structures, the raw materials must be denser such as bituminous coal. Within the GAC particle, these outstretched flat graphite platelets are placed indiscriminately to provide the extensive internal structure needed for adsorption to occur. AC treatment effectively removes chlorine-phenols chlorinated hydrocarbons, surfactants' color, and odor-producing substances from textile effluent. It can also substantially reduce harmful contaminants including heavy metals such as copper, lead, mercury, disinfection by-products. AC filters are positively charged that attract and trap many contaminants and prevent them from passing through the pores in AC filters can become plugged over time making them ineffective to remove any more contaminants. Thermal regeneration is generally used to reactivate the carbon by which the organics are evaporated out. In series when one is exhausted can be removed from service or column can be replaced. In parallel columns, both the beds become saturated simultaneously and both have to be replaced. AC can be added directly to the effluent, which eliminates needs of carbon adsorption beds. DuPont developed a process in which powdered AC is directly added to the aeration tanks [54–56].

Granular activated carbon
Granular activated carbon (GAC) is a particular formulation of AC, or activated charcoal. In ancient India, for drinking water filtration AC was used, and as a multi-use purifier in ancient Egypt. In modern day, in the early nineteenth century, it was introduced to Europe's sugar refining industry. Organic materials with high carbon contents such as wood, lignite, and coal are used in making GAC. The essential feature that distinguishes GAC to PAC is its particle size. Depending on the material used and manufacturing process, GAC typically has a diameter ranging between 1.2 and 1.6 mm and an apparent density ranging between 25 and 31 lb/ft^3. The bed density is about 10% less than the apparent density and is used to determine the amount of GAC required to fill a given size filter. To promote stratification after backwashing and minimize desorption and premature breakthrough, the uniformity coefficient of GAC is quite large, typically about 1.9, which results in mixing AC particles and adsorbed compounds with AC particles. A process is known as adsorption work through the preparations of AC. In absorption, like as a sponge soaking up water compounds are evenly distributed throughout the absorbent product. In creating a film, however in adsorption, the compounds bind only to the surface molecules and carbon molecules are naturally attractive. Therefore, they actively seek to bond to other molecules [57, 58].

Working
For locating a GAC treatment, there are two most common option units in water treatment plants, such as:

a. Post-filtration-adsorption—after the conventional filtration process (post-filter contactors or adsorbers), the GAC unit is located,
b. Filtration-adsorption—GAC replaces some or all of the filter media in a granular media filter.

Examples of these configurations are shown in Figs. 4 and 5, respectively. The GAC contactor receives the highest quality water in post-filtration applications, and thus, it has the only objective to remove the dissolved organic compounds. Adsorbers which are backwashed are usually unnecessary unless excessive biological growth occurs. By providing longer contact times than filter-adsorbers ensure the stipulates of most flexibility for handling GAC and for designing specific adsorption conditions.

The filter-adsorber configuration uses the GAC for turbidity and solid removal from textile effluent, and biological stabilization in addition to dissolved organics

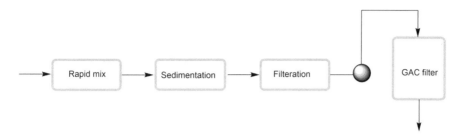

Fig. 4 Post-filtration-adsorption of granular activated carbon process

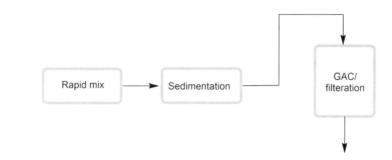

Fig. 5 Filtration adsorption of granular activated carbon process

removal. By replacing all or a portion of the granular media with GAC, the existing rapid sand filters can often be retrofitted for filtration-adsorption. Capital costs are reduced significantly by retrofitting existing high-rate granular media filters since no additional filter boxes, under-drains, and backwashing systems may be needed. Nevertheless, filter-adsorbers backwashed more frequently than post-filter-adsorbers (filter-adsorber units are backwashed about as frequently as conventional high-rate granular filters) since it has shorter filter run times and in addition, due to increased backwashing filter-adsorbers may incur greater carbon losses and may cost more to operate because carbon usage is less effective. In order to determine the required GAC contactor volume, the primary factors are the (i) breakthrough, (ii) empty bed contact time (EBCT), and (iii) design flow rate. The time when the concentration of a contaminant in the effluent of the GAC unit exceeds the treatment requirement is the breakthrough time. The GAC is exhausted and must be replaced/regenerated when the GAC effluent concentration is greater than the performance standard for over three consecutive days which is considered as a rule of thumb. The EBCT is calculated as the ratio of empty bed volume to that of flow rate through the carbon. By increasing the bed volume or reducing the flow rate through the filter, the longer EBCTs can be achieved. The amount of carbon to be contained in the adsorption units is defined by the EBCT and the design flow rate. Breakthrough and reducing the GAC replacement/regeneration frequency can be delayed by longer EBCT. Formerly the optimum EBCT is established, the carbon depth and adsorber volume can be determined. Typical EBCTs for water treatment applications range between 5 and 25 m. For GAC filters, the surface loading rate is the flow rate through a given area of GAC filter bed which is expressed in units of gpm/ft^3. Surface loading rates for GAC filters typically range between 2 and 10 gpm/ft^3. Highly adsorbable compounds which can be used in high surface loading rates are targeted for removal. The surface loading rate is not important when the mass transfer is controlled by the rate of adsorption as is the case for less-adsorbable compounds. The carbon usage rate (CUR) is used to determine the rate at which carbon will be consumed and how often carbon must be replaced/regenerated. By increasing, contact times improve with carbon treatment effectiveness. The percentage of carbon is increased by the deeper beds that are exhausted at breakthrough. After carefully evaluating the capital investment, the optimum bed depth and volume are typically selected and operating costs associated with reactivation frequency and contactor construction costs [59–63].

GAC contractors can be organized as either,

- Downflow fixed beds,
- Upflow fixed or expanded beds, or
- Pulsed beds, with single or multiple adsorbers operated in series or in parallel.

Each unit is connected in series with downflow fixed beds in series with the first adsorber in order to acquire the highest contaminant loading and the last unit receiving the lightest contaminant load. From the first unit, carbon is removed for reactivation, with the next adsorber becoming the lead unit. For parallel downflow fixed beds, each unit acquires the same flow and contaminant load. In order to maximize the carbon usage, multiple contactors are frequently operated in a parallel-staggered mode in which each contactor is at a different stage of carbon exhaustion. Since effluent from each contactor is integrated, individual contactors can be controlled beyond breakthrough such that the blended flow still meets the treatment goal. To remove the SS, periodic bed expansion is permitted by the up flow expanded beds and allows using smaller carbon particles without significantly increasing the head loss. Removal of spent carbon occurs from the bottom of the bed while fresh carbon is added at the top without system shutdown in the pulsed bed adsorbers. A pulsed bed cannot be entirely washed-out, which prevents contaminant breakthrough in the effluent [59–63].

3.3.2 Evaporation

Evaporation is an important tertiary treatment process; in textile industry, evaporation is used to dehydrate the concentrated solutions (e.g., caustic recovery from mercerizer) and separate dissolved solids, dyes, water from dye bath, and the rejected water from nanofiltration and RO process can be separated through evaporation through vaporizing and condensing of pure water. In case of effluent treatment, the permeate from RO as well as condensate water from evaporator results in pure distilled/demineralized water which can reuse in various textile processing or boilers [64–66]. Textile effluent is usually evaporated by the following two options.

a. Solar evaporation
b. Mechanical evaporation

Solar evaporation
The saline effluent (concentrate dye bath, first wash, rejected from RO, nanofiltration) is let into the large surface area in the open tanks, called as solar ponds (0.5–0.75 m height). For continuous operation and retention, multiples of solar ponds are constructed which gets filled up on a daily basis. The wind flow and sunlight available during daytime evaporate the surface water, and vapors are taken into the atmosphere (Fig. 6). The dissolved solids or concentrated dehydrated mass gets settled at the bottom and is periodically removed. The sludge is usually taken for secured landfilling storage.

Advantages

- Simple construction
- No energy costs involved
- No maintenance required
- No skilled technicians are required

Fig. 6 Solar evaporation plant used in small-scale textile coloration units in Tirupur, India

Table 11 Operating conditions for typical multiple evaporation plant (adopted and modified from [37])

Parameter	Stage I	Stage II	Stage III	Stage IV
Temperature °C	82	72	62	55
Length of tube (meters)	6	6	4.5	4.5
Diameter of tube (meters)	90	90	45	45

Disadvantages

- Very slow rate of evaporation
- Ineffective during raining seasons, low wind flows, cold climate, etc.
- Calls for larger surface area
- Chances of overflow contamination into groundwater in case of rain
- The release of water vapor into atmosphere leads to lack of recovery of pure water back for recycling.

Mechanical evaporation

Mechanical evaporators are heated by steam condensing on metallic, cylindrical tubes. As steam condensates on the outside of the tube, the heat of condensation causes saline water slurry on the inside of the tubes of the cylinder to get boiled. A mechanical evaporation is a device used for desalination or separation of pure liquid from admixtures, dehydration of dilute solution for concentration, separation of pure water from effluent slurries. In textile effluent treatment, evaporators are used for direct evaporation or stagewise concentrating saline effluent called as multiple effect evaporation plant (MEEP) (dye bath, first wash, RO, or nanorejects) followed by fluid crystallizer to separate solid and water. Evaporators can be batch or continuous operation, and the configuration can be vertical or horizontal. MEEP is a vertical cell in which stainless steel tube could be mounted; however, some of the evaporators are working with falling liquid film theory. The tube is heated with through steam, in generally the concentrated liquid is collected in the bottom of the evaporator, which is sent to next stage of the evaporator, where the freshwater is added to the first stage of the evaporator. Meanwhile, steam could be applied to the first stage of the evaporator; the next stage of evaporator gets steam from previous one. During evaporation, the liquid is re-circulated till it gets desired concentration [67, 68]. The operating parameters for MEEP are given in Table 11,

3.3.3 Electrocoagulation (EC)

Electrocoagulation was first proposed by Vik et al. [69]. It was designed for London sewage treatment plant. In 1909, Harries J. T received the USA patent for wastewater treatment by electrolysis using sacrificial electrodes made through aluminum and iron. Later, Matteson et al. [70] described the "Electro Coagulator" which can produce aluminum hydroxide from hydroxyl ions by using aluminum electrodes. Electrocoagulation (EC) is a well-established technology for the sustainable treatment of wastewater without the addition of chemicals such as Ferric, polyaluminum chloride (PAC). It is an electrochemical process which removes the SS, colloidal material, and metals, as well as other dissolved solids from the wastewater. Literally, the process called as "electrolysis" which breaks the substance by using of electricity; here direct current (DC) can be applied through to metal electrodes, and it starts reacting on an atomic level at the electrode–water interface from neutral, elemental state to charged state [71]. Thereafter, the charged metal atoms left the plate and enter into the water. Therefore, the electricity is applied to the electrodes to create the charged metal coagulant which bonds with dissolved contaminants, precipitates out of solution, organics adsorb to the precipitated solids, and solids agglomerate and separate, in another word electrical current destabilizes the SS allowing them to precipitate. Generally, the electrode is made from mild steel or aluminum which consume electrochemical reaction, therefore it should be replaced periodically [72]. The advantage of this process is that it generates less sludge than another conventional coagulation process such as a chemical method. Further, it is easy to dewatering the sludge, which drastically reduces the disposal cost. Also, it removes wide range of pollutant which includes heavy metals, suspended/colloidal solids, bacteria, BOD, COD, hydrocarbons, pesticides, and herbicides. Typical removal efficiency is around 95% and plus, and this process is good pre-treatment to membrane technologies where the high quality of water is required to reuse (Table 12). Presently, this process has gained attention due to its ability to treat large volume and for its low cost [73–76].

Table 12 Electrocoagulation treatment with a different type of dyes

Dye	Current or current density	Anode cathode	Removal efficiency (%)	References
Reactive orange 84	130 A/m^2	SS–SS, Fe–Fe	66, 76	Yuksel [77]
Reactive Black B, Orange 3R, Yellow GR	0.0625 A/cm^2	Al–Al	98	Khandegar [78]
Reactive Black 5	7.5 mA/cm^2	Fe–SS	90	Patel [79]
Remazol Red RB 133	15 mA/cm^2	Al–Al	92	Can OT [80]
Direct Red 23	30 A/m^2	Fe–Fe, Al–Al	>95	Phalakornkule [81]
Acid Red 14	80 A/m^2	Fe-St 304	93	Daneshvar [82]
Acid Orange 7	5 mA/cm^2	Boran-doped diamond–copper foil	98	Fernandes [83]
Disperse Red	20.8 mA/cm^2	Al–Al	95	Merzouk [84]

Mechanism of Electrolysis

In this process, coagulant can be generated due to the electrochemical reactions which produce the ions and help to remove the various kinds of pollutants either by chemical reaction and precipitation or by causing the colloidal materials to coalesce. The main processes are occurring during the electrochemical reaction is electrolysis which held in the electrodes, formation of coagulants in the aqueous phase, adsorption of soluble or colloidal pollutants on coagulants, and removal by sedimentation and floatation. The main chemical reaction in the electrodes is as follows;

$$Al \rightarrow Al^{3+} + 3e^- (\text{at anode})$$
$$3H_2O + 3e^- \rightarrow 3$$
$$2H_2 + 3OH^- (\text{at cathode})$$

The flocculation and coagulation in the EC technique have the advantage that it can be possible to remove the small colloidal particles, whereas in the conventional method it could not be. In addition, it reduces the sludge levels and also it is a very simple process, therefore easy to operate. The simple principle of electrolysis is shown in Figs. 7 and 8.

Mode of EC operation

The mode of EC operation can be divided into two categories, which is batch and continuous processes. A continuous process of EC can be operated under steady-state conditions, particularly with respect to the fixed pollutant concentration and effluent flow rate; however, it is perfectly matched with bulk quantity (i.e., for industrial) of effluent, whereas batch reactors are suited to laboratory and pilot plant scale applications. In a continuous process, it is very easy to control the operational parameters; therefore, it provides the better efficiency. The batch process is mostly time-dependent, and the coagulant is continuously generated by the reactor with the dissolution of anode. Therefore, anode can hydrolyze the pollutants, so the concentration of pollutant, coagulant, and pH can change time to time. The schematic representation of batch and continuous EC processes is given in Fig. 9.

Impact of Various Operating Parameters

However, the efficiency of EC process can be depended on the various operational parameters such as conductivity of the solution, arrangement of electrode, electrode shape, type of power supply, pH of the solution, current density (CD), distance between the electrodes, agitation speed, electrolysis time, initial pollutant concentration, retention time, and passivation of the electrode [86].

Solution conductivity

In the electrolysis, solution conductivity is very important; however, removal efficiency and operational cost of EC process are based on the solution

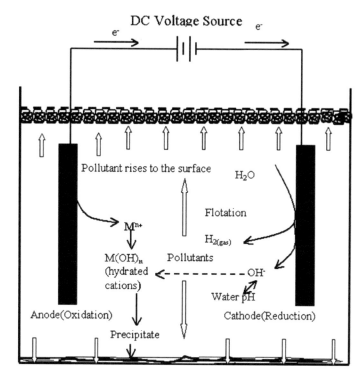

Fig. 7 Simple principles of electrocoagulation process (reprinted from [85], with kind permission of Elsevier publications)

Fig. 8 Laboratory-scale electrocoagulation process

conductivity. In general, the solution must have minimum conductivity, and then only electric current could be flowing in the solution. However, the addition of electrolyte (sodium chloride or sodium sulfate) can regulate the conductivity of low-conductivity wastewater. Perhaps, the energy consumption could be less in case of higher conductivity solution [87].

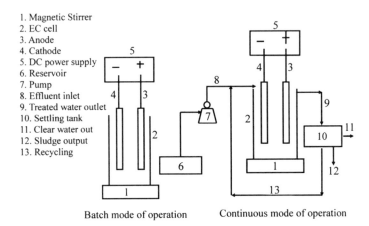

1. Magnetic Stirrer
2. EC cell
3. Anode
4. Cathode
5. DC power supply
6. Reservoir
7. Pump
8. Effluent inlet
9. Treated water outlet
10. Settling tank
11. Clear water out
12. Sludge output
13. Recycling

Batch mode of operation   Continuous mode of operation

Fig. 9 Schematic diagram of batch and continuous modes of operation (reprinted from [86], with the kind permission of Elsevier publications)

Arrangement of electrodes

In EC process, the operational cost linearly depends on the connection mode of the electrode; however, it is confirmed by a previous study [88, 89]. Nasrullah et al. [89] studied the different arrangement of electrodes and their efficiency. Therefore, authors used to keep the electrodes in parallel, series, monopolar, and bipolar aluminum electrodes to find the optimum results. The arrangements (i.e., monopolar parallel (MP-P), monopolar series (MP-S), and bipolar (BP)) of the electrode can be described in the Fig. 10. The anode and cathodes are arranged in parallel order with respect to the MP-P setup; here the electric connection can be divided into the electrodes. In MP-S, electrodes are connected to the power source forming anode and cathode, while a pair of the inner electrodes (also called as sacrificial electrodes) is connected to each other without any connection to the power source. In BP configuration, there are two sacrificial electrodes which are placed in between the two parallel electrodes without any connection. Also, authors used different electrode orientation, namely vertical, horizontal orientation with anode at the bottom and horizontal orientation with anode on top; among these, vertical orientation provides higher percentage of COD removal, and it reached 59% with respect to the 120 m, whereas horizontal orientation with anode at the bottom shows only 40% of same time.

The results for the different arrangement of electrodes, MP-S showed 65, 65, and 60% for COD, BOD, and SS, respectively, in 120 m, whereas BP shows 56, 56, and 51%, which is lowest treatment efficiency toward to various pollution parameters (Fig. 11). It is clearly shown that the highest removal efficiency can get through serial connection than parallel connection. This is due to the higher consumption of anode material; therefore, the efficiency of pollutant treatment purely depends on the anodic dissolution in the system, where higher dissolution of anode makes more coagulants to the system, which assures more treatment efficiency.

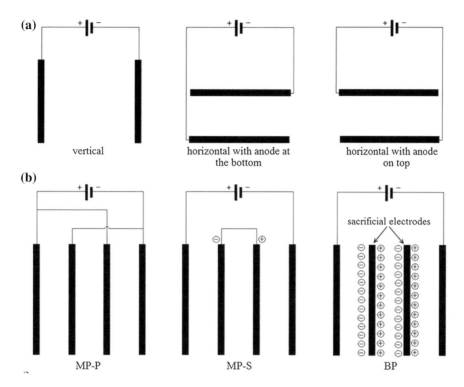

Fig. 10 Arrangement of electrode in EC process: **a** orientations, **b** arrangements (reprinted from [89], with the kind permission of Elsevier publications)

Rotating anode

Ahmed Samir et al. [90] studied the overall efficiency of novel rotating electrodes with respect to the various speed and different CD on textile effluent (Imperon Violet KB (CAS #: 6358-46-9) for the fabric dyeing process.), schematic representation of rotating electrode are shown in Fig. 12.

The overall results of total removal efficiency are shown in Fig. 13, COD removal efficiency is 90% when the anode was static with no rotational speed (CD = 8 mA/cm^2) for the time period of 30 m, meanwhile the efficiency can be increased to 91% by increasing the rotational speed of electrode by 75 rpm and reduced the CD = 4 mA/cm^2 for the time period of 10 m; similarly, the efficiency can be increased to 95% by increasing the rotational speed of 150 rpm with the same CD, and during this period, the formation of Al(OH)$_3$ flocs which connected to other pollutants makes precipitation easier. But when the electrode speed is increased to 250 rpm, there is no COD removal efficiency can be achieved, which is due to the extremely high speed of the electrode cannot make the electrolysis process. In general, COD removal efficiency was increased by increasing the CD. The initial color and TSS were decreased within 10 m with the rotational speed of

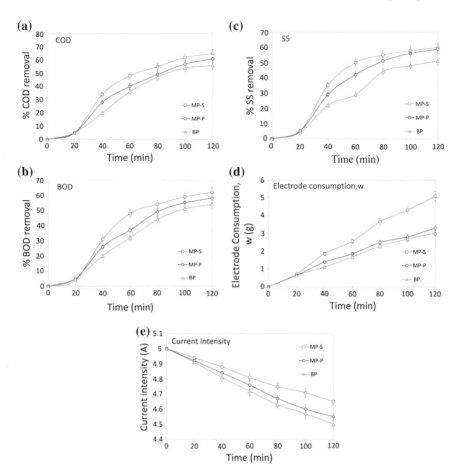

Fig. 11 Effect of MP-S, MP-P, BP as electrode arrangements on **a** COD removal, **b** BOD removal, **c** SS, **d** anodic dissolution, and **e** current intensity changes during 120 m (reprinted from [89], with the kind permission of Elsevier publications)

150 rpm, and the results were significantly varied with respect to the CD; however, very good color removal can be obtained by application of 8 mA/cm^2.

Shapes of the electrode

In EC process, the shape of electrodes is directly influenced on the pollutant removal efficiency. Yosuke Kuroda et al. [91] studied on the different shape of electrodes with respect to their pollutant removal efficiency and concluded that the punched hole type of electrode shows higher removal efficiency than the electrode without punched holes (Fig. 14). However, there are very few studies conducted on and reported in this field [92]. Kuroda et al. [91] reported in their studies the higher discharge current for the electrode with punched holes. However, the higher discharge current is directly influenced by higher pollutant removal efficiency. Perhaps, it still required higher number of studies to establish this study.

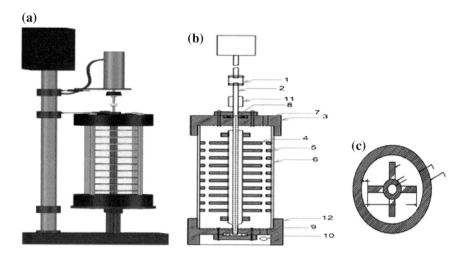

Fig. 12 a Schematic diagram of EC rotated anode system. **b** Details of EC rotated anode reactor: 1. motor variable speed; 2. stainless steel shaft; 3. upper Teflon flange cover; 4. Al rods of impellers anode; 5. Al rings of cathode; 6. Perspex reactor; 7. upper ports; 8. ball bearing; 9. thrust bearing; 10. lower port; 11. zoom coupling; 12. lower Teflon flange cover. **c** Top view of impellers anode with ring cathode (reprinted from [90], with kind permission of Elsevier publications)

Type of power supply

In the EC process, the electrolysis reaction can be performed through electrode, and it required electric voltage; the majority of the research has used DC and few are in alternative current [93, 94]. However, the use of DC leads to corrosion formation on the electrode due to the metal oxidation during the EC process, and also the oxidation layer can be formed in the cathode and followed by it reduce the current flow between the cathode and anode, results are reduced pollutant removal efficiency. Subramaniyan et al. [95] have studied the effect of alternative current and DC on pollutant removal efficiency of cadmium from water by using the aluminum alloy as anode and cathode. They obtained improved results with respect to the alternative current (97.5%) than DC (96.2%) (Fig. 15), at 0.2 A/dm^2 of CD with pH of 7, the results provide the information that alternative current will reduce the formation of corrosion in the electrode, which resulted in increased removal efficiency.

Eyvaz [94] analyzed the effects of alternating current electrocoagulation on textile wastewater for dye removal, and wastewater containing disperse yellow 241 (DY) and reactive yellow 135 (RY) dyes. They obtained the results of dye removal efficiency which are plotted in Fig. 16; in overall the reactive dye (RY) shows lower removal efficiency than disperse dye (DY) in both the power (i.e., alternative current and DC). However, after 100 m, alternating current power supply shows higher removal efficiency for both the dyes, which means that increasing the operation time can increase the accumulation of dye residue causing passivization on cathode material.

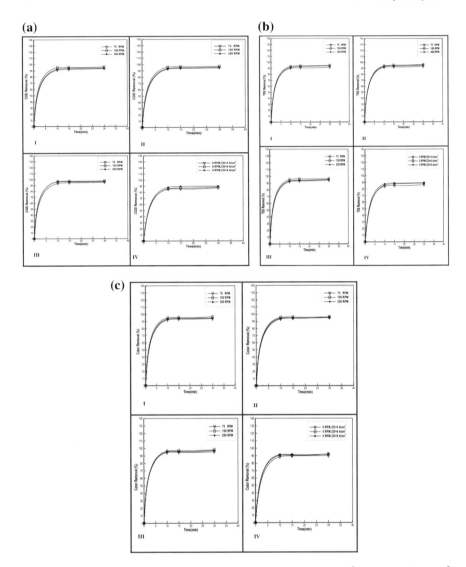

Fig. 13 a Effect of rotational speed on COD removal, I. CD = 4 mA/cm²; II. CD = 6 mA/cm²; III. CD = 8 mA/cm²; IV. Zero rotational speed. **b** Effect of rotational speed on TSS removal, I. CD = 4 mA/cm²; II. CD = 6 mA/cm²; III. CD = 8 mA/cm²; IV. Zero rotational speed. **c** Effect of rotational speed on color removal, I. CD = 4 mA/cm²; II. CD = 6 mA/cm²; III. CD = 8 mA/cm²; IV. Zero rotational speed (reprinted from [90], with the kind permission of Elsevier publications)

pH of the solution

In the process of electrocoagulation, the important operational parameter is the pH of the solution. For a particular pollutant, the efficiency of removing the maximum pollutant is obtained at an optimum solution pH. At a particular pH, the

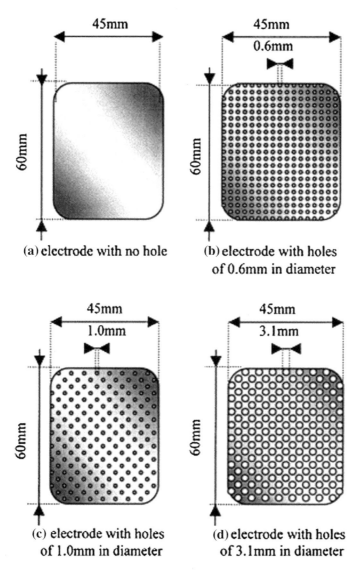

45mm

60mm

(a) electrode with no hole

45mm
0.6mm

60mm

(b) electrode with holes
of 0.6mm in diameter

45mm
1.0mm

60mm

(c) electrode with holes
of 1.0mm in diameter

45mm
3.1mm

60mm

(d) electrode with holes
of 3.1mm in diameter

Fig. 14 Different electrode shapes used in EC process (reprinted from [91], with the kind permission of Elsevier publications)

precipitation of a pollutant begins. The pH of the solution from the optimum pH either decreases or increases according to the decreasing the pollutant removal efficiency. The removal of hexavalent chromium from synthetic solution was studied by Verma et al. [96] using electrocoagulation and found that the pH of the solution has a significant effect on the Cr(VI) removal efficiency. At different pH of the synthetic solution, they performed the experiments and acquired the maximum chromium removal efficiency at the pH 4. They concluded that the synthetic

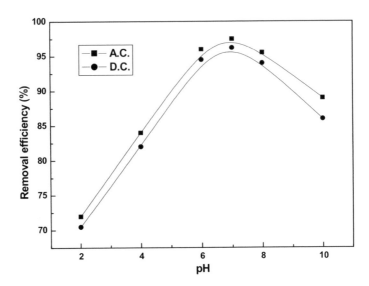

Fig. 15 Effect of pH of the electrolyte on the removal of cadmium. Conditions: an electrolyte concentration of 20 mg/l; a current density of 0.2 A/dm^2; temperature of 303 K (reprinted from [95], with the kind permission of Elsevier publications)

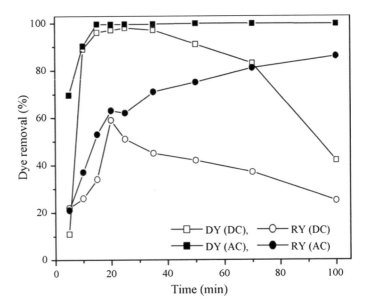

Fig. 16 Effects of power supply type on dye removal in view of operation time (reprinted from [94], with the kind permission of Elsevier publications)

solution pH increases after the EC process with an increase in the electrolysis time due to the generation of OH in the EC process.

Fajardo et al. [88] have studied the influence of initial pH, applied CD and initial dye concentration on the efficiency of EC process with respect to textile wastewater. To asses the pH influence, they used different pH (3, 6, and 9) by adjusting with H_2SO_4 or NaOH. The pollutant removal efficiency was calculated with respect to the time; there is a significant improvement in COD removal when the initial pH of pollutant at 3, meanwhile, pH 6 and 9 could not show much difference within 30 m, thereafter all pH shows almost similar results. However, after 120 m, pH 9 will give a better result than another pollutant. In the case of color removal, before 60 m, there is a huge difference in the efficiency and thereafter the results are almost nearer, which means there is no significant difference (Fig. 17).

Current density
In electrocoagulation, the important parameter, namely CD, determines the coagulant dosage rate, bubble production rate, size and growth of the flocs, and in turn, it affects the efficiency of the electrocoagulation. As the CD increases, the anode dissolution rate increases. As a result, it increases the number of metal hydroxide flocs resulting in the increase in pollutant removal efficiency. When the CD increases above the optimum CD, it does not result in an increase in the pollutant removal efficiency due to a sufficient number of metal hydroxide flocs are procurable for the sedimentation of the pollutant. Fajardo et al. [88] also studied the impact of applied CD (i.e., authors used 4 and 16 mA/cm^2) on the color removal; the results are plotted in Fig. 18. However, the highest color removal efficiency of 97.1% was found with respect to the 16 mA/cm^2 at the time period of 90 m. Conversely, 360 m was necessary to achieve similar decolorization percentage when 4 mA/cm^2 was applied. Apart from that, the COD removal efficiency is not showing the significant difference with respect to the applied CD.

Distance between the electrodes
The electrostatic field depends on the distance of anode and cathode; therefore, it plays a significant role in EC process. Maximum pollutant removal efficiency could be found with at optimum distance of anode and cathode. Perhaps, pollutant removal efficiency is less when the electrode distance is minimum; it is due to the metal hydroxides generated and followed by flocs, which refuse to sediment. Daneshvar et al. [82] studied the effect of interelectrode distance and their pollutant removal efficiency of Acid Red 14 dye. The result is shown in Fig. 19; color removal efficiency can be increased with increasing the distance of interelectrode (anode and cathode). This is due to the electrostatic effects between the interelectrode during the EC process; therefore, the distance increases, movement of produced ions could be slow in the action and possible to aggregate and produce the flocs.

Effect of agitation speed
In order to maintain uniform conditions, the agitation is helpful and it avoids the formation of the concentration gradient in the electrolysis cell. Further, for the

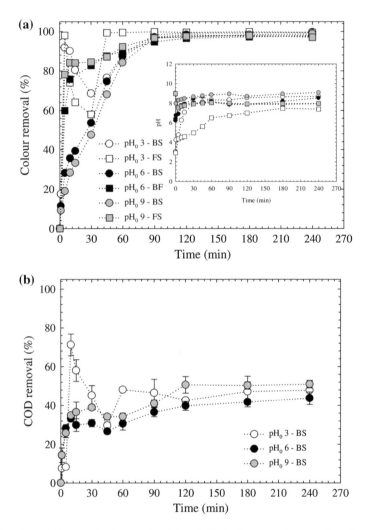

Fig. 17 Effect of the initial pH on color (**a**) and COD (**b**) removals over time. Inset **a**: pH evolution over time. Operating conditions: $j = 16$ mA/cm^2, 100 mg/L of Reactive Black 5, stirring rate = 800 rpm (batch system), and flow rate = 160 l/h (flow system) batch-stirred system (BS) versus recirculation flow system (FS) (reprinted from [88], with the kind permission of Elsevier publications)

movement of the generated ions, the agitation in the electrolysis cell imparts velocity. In order to increase the pollutant removal efficiency, the agitation speed is increased up to the optimum agitation speed. For a particular electrolysis time, the flocs are formed much earlier which consequently increase the pollutant removal efficiency due to the fact that with an increase in the mobility of the generated ions. But there is a decrease in the pollutant removal efficiency when it is increased

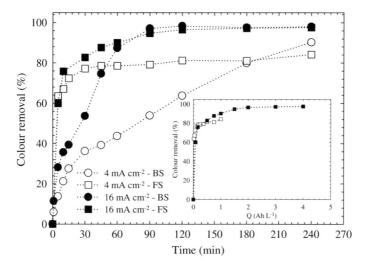

Fig. 18 Effect of the current density on color removal over time, for batch and flow systems. Inset: Color removal as a function of the applied charge at different current densities for the flow system. Operating conditions: pH = 6, 100 mg/L of Reactive Black 5, stirring rate = 800 rpm (batch system), and flow rate = 160 l/h (flow system) (reprinted from [88], with the kind permission of Elsevier publications)

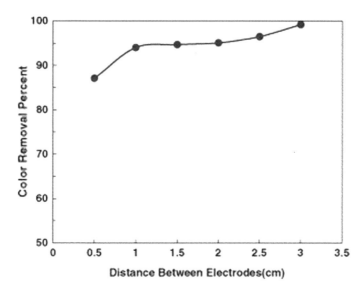

Fig. 19 Effect of interelectrode distance on the efficiency of color removal from a solution (reprinted from [82], with the kind permission of Elsevier publications)

further in the agitation speed beyond the optimum value, as the flocs get degraded by collision with each other due to high agitation speed [97].

Electrolysis time

One of the functions of electrolysis time is the pollutant removal efficiency. With an increase in the electrolysis time, the pollutant removal efficiency increases. But when it goes on the far side of the optimum electrolysis time, the pollutant removal efficiency remains constant and does not increase with an increase in the electrolysis time. By desolating the anode, the metal hydroxides are formed. When a CD is fixed, the number of generated metal hydroxide increases with an increase in the electrolysis time. There is an increase in the generation of flocs for a longer electrolysis time, resulting in an increase in the pollutant removal efficiency. When the electrolysis time goes beyond the optimum electrolysis time, the pollutant removal efficiency does not increase as sufficient numbers of flocs are available for the removal of the pollutant. Aoudj et al. [98] studied the different operational parameters of EC process for Direct Red 81 containing effluent. In this study, they studied the reaction time, results are plotted in Fig. 20, and the results are increased from 52.5 to 98.28% by increasing the time from 10 to 60 m, since other parameters are constant (CD of 2.5 mA/cm^2, initial dye concentration, 50 mg/L and initial pH 6). However, the reaction has linear relationship with pollutant removal efficiency of EC process. This is due to the anodic electrodissolution which led to release the coagulating process. In general, the dye removal efficiency depends on the concentration of metal ions which can be produced by electrodes. Therefore, the concentration of metal ions is directly proportional to the electrolysis time.

Initial concentration of pollutant

For a constant CD, there is an increase in the initial concentration of the pollutant as the pollutant removal efficiency decreases. Due to the insufficiency of the number of metal hydroxide flocs formed in order to sediment the greater number of pollutant molecules at higher initial pollutant concentrations [99]. Therefore, lower the initial

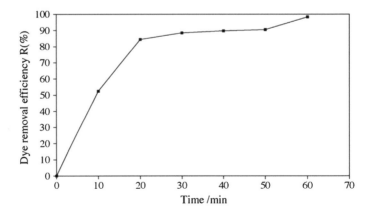

Fig. 20 Effect of time of electrolysis on the removal efficiency of Direct red 81 (reprinted from [98], with kind permission of Elsevier publications)

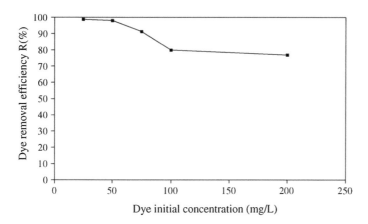

Fig. 21 Effect of initial dye concentration on the removal efficiency of Direct Red 81 (reprinted from [98], with the kind permission of Elsevier publications)

dye concentration makes better decolorization efficiency. Aoudj et al. [98] obtained the results of initial dye concentration with respect to dye removal efficiency, and when the initial dye concentration is increased from 25 to 200 mg/L, the dye removal efficiency could be decreased from 98 to 76% (Fig. 21).

Retention time

Once the electrocoagulation process for a particular electrolysis time is completed, the solution is kept for a fixed period (retention time) in order to allow settling of the coagulated species. The removal efficiency of pollutant increases as the retention time increases. Due to the increase in retention time, all coagulated species settle down easily to give clear supernatant liquid and the sludge. The optimum retention time resulting in the reduction of pollutant removal efficiency as the adsorbed pollutant desorbs back into the solution [82]

The advantages of electrocoagulation as compared to chemical coagulation are as follows:

- Compared to the chemical coagulation, EC requires no addition of chemicals and also provides better removal capabilities for the same species.
- As the chemical coagulation cannot remove many species but for EC removes.
- In order to lower the sludge disposal cost, EC produces less sludge.
- EC sludge can be utilized as a soil additive which is more readily filterable
- EC sludge contains metal oxides which makes it pass the leachability test.
- EC technique process can be started by turning on the switch which needs only a minimal startup time.

Some of the limitations of the electrochemical coagulation are as follows:

- Periodically, the sacrificial anodes need to be replaced.

- Depending on reactor design, the electrocoagulation requires a minimum solution conductivity in order to limit its use with effluent containing low dissolved solids.
- In case of the removal of organic compounds, there is a possibility of formation of toxic chlorinated organic compounds from effluent containing chlorides.
- On the cathode, an impermeable oxide film may be formed which may provide resistance to the flow of electric current. However, changing the polarity and cleaning of the electrodes periodically may reduce this interference.
- There is an increase in the operational cost of EC due to the high cost of electricity.

3.3.4 Advanced Oxidation Process (AOPS)

The main goal of AOP is to produce and use the hydroxyl free radical (HO), and it is a strong oxidant and destroys the compound; however, it is not possible by the conventional oxidant [100]. Generally, we use conventional oxidation process to treat the drinking water; nevertheless, this process may not destroy all kind of toxins. Every oxidizing agent has the different oxidation potentials, which are listed in Table 13. AOP is characterized to produce the OH radicals, and it can specifically attack, this is the biggest versatility, and also it can produce different methods which are given in Table 14.

Ozonation
Ozone is a strong reactive gas with oxidation properties; generally, it is unstable, explosive, and easily recognized by its smell. However, it is toxic in nature even very less concentration, the density of ozone is ~ 1.5 times higher than the atmospheric air [102–105]. Ozone has very good solubility in water than the oxygen. However, ozone can be used to treat the drinking water to kill the microorganisms like bacteria and other living things [106]. The mass transfer of ozone from gas-phase to liquid-phase is a limiting step. The driving force and

Table 13 Conventional oxidizing agents and its oxidizing potential

Oxidizing agent	EOP (V)	EOP related to chlorine
Fluorine	3.06	2.25
Hydroxyl radicals	2.8	2.05
Atomic oxygen	2.42	1.78
O_3	2.08	1.52
H_2O_2	1.78	1.3
Hypochlorite	1.49	1.1
Chlorine	1.36	1
Chlorine dioxide	1.27	0.93
Oxygen (molecular)	1.23	0.9

Note EOP Electrochemical oxidation potential

Table 14 Different methods to generate AOP with respect to the wastewater treatment (reprinted from [101], with the kind permission of Springer publications)

AOP types	Oxidant for advanced oxidation	Other occurring mechanisms
O_3	OH^-	Direct O_3 oxidation
O_3/H_2O_2	OH^-	Direct O_3 oxidation H_2O_2 oxidation
O_3/UV	OH^-	UV photolysis
UV/TiO_2	OH^-	UV photolysis
UV/H_2O_2	OH^-	UV photolysis H_2O_2 oxidation
Fenton reaction	OH^-	Iron coagulation Iron sludge-induced adsorption
Photo-Fenton reaction	OH^-	Iron coagulation Iron sludge-induced adsorption
Ultrasonic irradiation	OH^-	Acoustic cavitations generate transient high temperature
Heat/persulfate	SO_4^-	Persulfate oxidation
UV/persulfate	SO_4^-	Persulfate oxidation UV photolysis
Fe (II)/persulfate	SO_4^-	Persulfate oxidation Iron coagulation Iron sludge-induced adsorption
OH^-/persulfate	SO_4^-/OH^-	Persulfate oxidation

efficiency of ozone treatment can be varied, and it is based on the nature of the liquid (i.e., concentration, pH of the effluent). A simple reliable method is used to inject ozone into the closed tank which contains the wastewater. The remaining ozone which is not consumed in the process would be destroyed in the off-gas with the help of thermal catalytic unit. There are two methods can produce the ozone, ultraviolet, and corona discharge. Generally, it is applied to wastewater through diffuser tubes or turbine mixers. The dosage of 2 mg/L itself shows significant impact on the pollutant removal efficiency [107, 108]. The previous literature shows that the degradation of azo dyes with ozone is quite limited; however, the ozone readily attacks the electron-rich molecules [105, 109, 110]. The oxidation potential of organic–inorganic matters by ozone is 2.07 V; however, in some cases, higher dose rate cannot help to convert organic maters to CO_2 and H_2O completely, particularly with textile dyes, which has the combination of effluent (surfactant, suspended matters, and other auxiliaries). Ozone treatment can produce the better pollutant removal efficiency when the parameters are optimized, such as temperature, pH, and applied ozone dose [111]. A simple decomposition mechanism of ozone in aqueous solution illustrated in the following equation [104],

$$O_3 + OH^- \rightarrow HO_2^- + O_2^{\cdot-} \tag{1}$$

$$O_3 + HO_2^{\cdot} \rightarrow OH^{\cdot} + 2O_2 \tag{2}$$

$$O_3 + OH^{\cdot} \rightarrow O_3^{\cdot} + OH^{\cdot} \tag{3}$$

$$O_3^{\cdot} \rightarrow O^{\cdot} + O_2 \tag{4}$$

$$O^{\cdot} + H^+ \rightarrow OH^{\cdot} \tag{5}$$

$$OH^{\cdot} + HO_2^{\cdot} \rightarrow H_2O + O_2 \tag{6}$$

Pazdzior et al. [112] studied the influence of ozone on textile effluents; therefore, they applied different ozone doses from 1.68 to and 0.14 g O_3 dm^{-3}; a diagram for the laboratory ozone treatment is shown in Fig. 22.

Ozone doses were showing significant improvement in the toxic reduction of textile dye (C.I Reactive yellow 186), the highest concentration of ozone dose will reduce $\sim 50\%$ of total toxic units than the control sample (sample without ozone treatment), and the results are shown in Fig. 23.

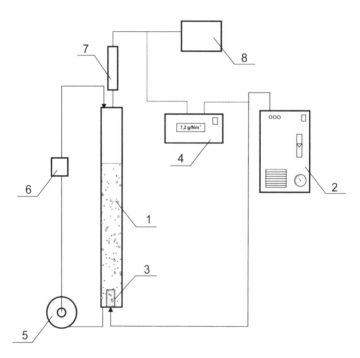

Fig. 22 Laboratory setup for ozonation process: 1 bubble column; 2 ozonator with oxygen concentrator; 3 gas diffuser; 4 gaseous ozone analyzer; 5 peristaltic pump; 6 cell for samples collecting; 7 scrubber filled with silica gel; 8 ozone destructor (reprinted from [112], with the kind permission of Elsevier publications)

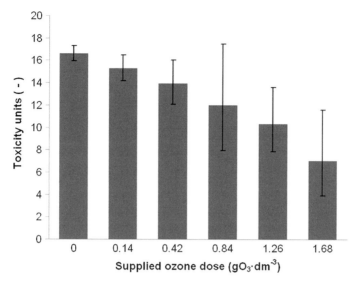

Fig. 23 Toxicity changes for different ozone doses (reprinted from [112], with the kind permission of Elsevier publications)

Fig. 24 Effect of pHs on percent UV absorption of Congo red at different ozonation time (reprinted from [113], with the kind permission of Springer publications)

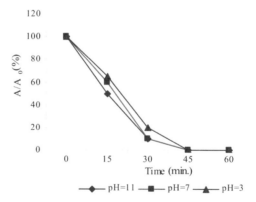

During the ozone treatment, pH of the effluent shows significant role in the pollutant removal [113]. Fig. 24 shows the color removal of Congo red by ozone treatment with different pH of the initial effluent. It was found that the deterioration of dyes could take place during the alkaline conditions. In general, the ozone process can improve the efficiency of biodegradation and biodegradability of the Congo red dye.

Advantages of ozonation

- Eliminates the conventional coagulation process,
- Ensures the safer process,

- Removes pesticides, organics, BOD, COD sewful and as well as decoloration,
- Minimizes secondary biological treatment load,
- Simple operation.

3.3.5 Ultrasonics

Chemical pollutants can be degraded by using ultrasonic technology, particularly the refractory organic pollutants in water. It also utilizes the combination of the characteristics of advanced oxidation technology, incineration, supercritical water oxidation, and other wastewater treatment technologies. Degradation speed is fast likewise the degradation conditions are also mild, its application is widely applied, and it can also be used individually or combined with other water treatment technologies. The sewage enters into the air vibration chamber after being added the selected flocculants in regulating tank which is the principle of this method. A part of organic matter in wastewater is changed into the small organic molecule by destructing its chemical bonds under the intense oscillations in nominal oscillation frequency. CODCr and the aniline concentration fall under the accelerating thermal motion of water molecules with the flocculants' flocculation rapidly companied with the color, in which it plays the role of reducing organic matter concentration in wastewater. At present, the ultrasonic technology in the research of water treatment has acquired the great achievements, but still, most of them are confined to laboratory research level [114–116]. Onat et al. [117] studied the decolorization of different reactive dyes and basic dye by the ultrasonic/microbial method. First, they treated effluents by ultrasonic then treated with microbial, ultrasonic decolorization systems are given in Fig. 25. Authors carried out the ultrasonic treatment by using ultrasonic homogenizer (Cole Parmer-CPX 600) at 20 kHz (600 W). The initial concentrations of dye solutions were 50 and 100 mg/L.

For decolorization, ultrasonic treatment was carried out for 5 h with 20 kHz; the results are significantly increased with increasing the treatment time with respect to

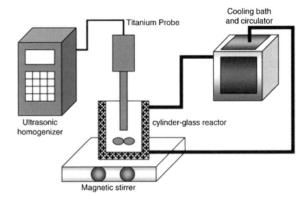

Fig. 25 Ultrasonic decolorization system (reprinted from [117], with the kind permission of Elsevier publications)

all three dyes. However, the decolorization is increased with decreasing the initial concentration of dyes. After 5 h, authors were obtained the results are 23, 28, and 16% for the initial concentration of effluent with 100 mg/L, whereas 31, 48, and 28% for 50 mg/L for reactive blue 4, reactive red 2, and basic yellow 2, respectively (Fig. 26).

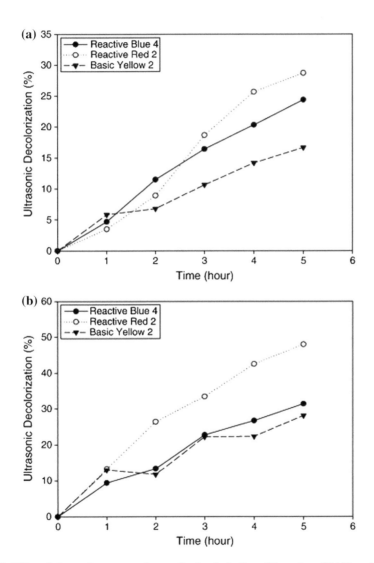

Fig. 26 Effect of ultrasonic treatment time on the decolorization of three dyes, 20 kHz. **a** 100 mg/L; and **b** 50 mg/L initial dye concentration (reprinted from [117], with the kind permission of Elsevier publications)

3.3.6 Sequencing Batch Reactor

The SBR is the modified technology of ASP. The advantages of SBR process are to control the process as well as the flexibility and alternatives of the process and its designs to achieve the latest effluent discharge standards. Initially, SBR technology was used mainly to sewage treatment, due to the technological, design feasibility and better process control that can be achieved by the modern technology makes to use SBR in much industrial wastewater treatments, including textile effluents. For the sewage treatment, SBR can save 60% of total operating expenses as compared to conventional ASP process, and it also achieves high pollutant removal efficiency. However, SBR process has disadvantages, the prime one is time consumption, it required more time, sometimes the process extends for 30 h, it is due to the slow nature of the SBR process, and it cannot manage the immediate reaction on the effluent; therefore, it is quite suitable for the small-scale industries. Generally, the operation of SBR contains five processes,

- Inflow,
- Reaction,
- Sedimentation,
- Outflow,
- Standby.

A simple schematic representation of SBR is described in the Figs. 27 and 28. It is a great resistance to the shock loading, the stored effluent can effectively resist the impact of water and other organic substance.

Sathian et al. [118] studied the effect of SBR treatment on pollutant removal efficiency for textile wastewater by simple modification of process parameters, the decolorization percentage having the good correlation with sludge retention time (SRT) as well as the air flow rate. Maximum color removal is found with the optimized process parameters.

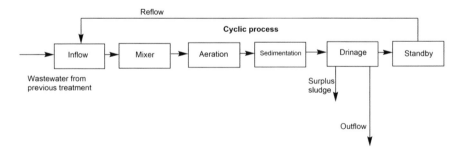

Fig. 27 Various reactions take place in SBR-based wastewater treatment process

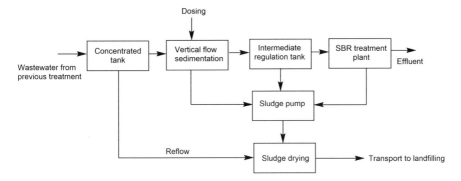

Fig. 28 SBR-based wastewater treatment process

3.3.7 Membrane Bioreactor (MBR)

Due to distinct advantages over conventional bioreactors, membrane bioreactor (MBR) technology has been used broadly for various industrial wastewater treatments. During the past decade, on evaluating the performance of MBR technology for textile wastewater, a significant number of research studies have been conducted. MBR is used in treatment of textile wastewater with a significant removal of contaminants has been investigated as a simple, reliable, and cost-effective process. However, membrane fouling is a major drawback in the operation of MBR, which leads to the decline in permeate flux and hence requires membrane cleaning. The lifespan of the membrane is eventually decreased. From MBR process, the high quality of the treated water is common to all commercial aerobic systems. A superior performance in the treatment is shown by MBR technology and operation of domestic and a wide spectrum of industrial wastewaters (including wastewater containing micropollutants) compared to other conventional treatment technologies [119]. Compared to aerobic MBR under similar operational conditions, Martin-Garcia et al. [120] reported that soluble microbial products in anaerobic MBR were 500% higher. This is one of the reasons anaerobic MBR has not been widely applied in wastewater treatment. Since the biomass characteristics are controlled by the SRT, it has an impact on membrane fouling. Generally, in increasing the mixed liquor suspended solids (MLSS) concentration is assisted by longer SRT in MBR system high concentration of MLSS are associated with the high viscosity of mixed liquor which significantly contributes to the membrane fouling has been reported by Xing et al. [121]. Figure 29 shows the schematic presentation of the laboratory-scale MBR. A bioreactor, divided into an aerobic reactor and anaerobic reactor, and an external submerged membrane module are included in the setup. An air pump controls the transmembrane pressure (TMP), pumping at a constant rate. Reversal of pressure was done in every 10 m and from the storage tank for 30 s, the membrane was back pulsed with permeate. An air compressor applied aeration along the membrane surface is used to prevent fouling on the membrane [122].

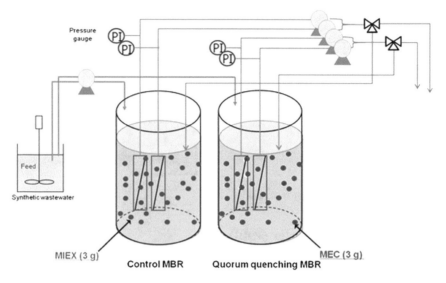

Fig. 29 Schematic presentation of the laboratory-scale MBR (reprinted from [123] Springer publications)

3.3.8 Photocatalysis

In this process, colored molecules occur as a result of initial absorption of radiation in wastewater that act as a photosensitizer. Compared to that of UV region, photosensitization is based on the utilization of longer wavelength, in order to degrade the organic compounds to CO_2, H_2O, and mineral acid as in the existence of a suitable conductor. For the decolorization of industrial effluents, UV light has been tested in combination with H_2O_2 or solid catalyst such as TiO_2. Due to cost, too slow, and little effective for potential full-scale application, the combination of UV/TiO_2 seems more promising than the UV/H_2O_2 process. During photocatalysis, electron–hole pairs are generated by TiO_2 when irradiated by the light of wavelength shorter than 380 nm. By direct hole transfer, the organic pollutants are thus oxidized or in most cases attacked by the OH radical formed in the irradiated TiO_2. For, TiO_2 photocatalysis, the ˙HO quantum yield has been determined to be 0.040 at 365 nm. During illumination with light having energy higher than the band gap energy of semiconductor-like titanium dioxide, zinc oxide, tungsten oxide can initiate decomposition, and often the complete mineralization of organic compounds. The process is carried out under ambient conditions which is the reason for the increased interest in the photocatalytic process which can be activated by UV light and does not require expensive oxidants with the catalyst being inexpensive and non-toxic [124–126].

An ideal treatment method for the removal of color, BOD, and COD from textile wastewaters was founded to be photocatalysis by Arslan et al. [127]. Maximum efficiency by photocatalytic oxidation was observed to be achieved at pH 4. For four non-biodegradable commercial azo dyes and one industrial wool textile wastewater using TiO_2 suspension irradiated with a medium pressure mercury lamp were investigated by photodegradation and biodegradability. In aqueous heterogeneous suspensions and in the solid state, the TiO_2/UV photocatalytic degradation of indigo and of indigo carmine has been investigated. It was observed that oxidation of dye, with almost complete mineralization of carbon, nitrogen, and sulfur heteroatoms into CO_2, NH_4, NO_3, and SO_4^{2-}, was achieved in addition to prompting removal of color. Modified titanium dioxide photocatalysis removed the color completely in relatively short time (60 m) compared to that of photocatalytic oxidation of dyes in water.

Hong et al. [128] investigated the effects of photocatalysis on the color removal activity and the growth of isolated photosynthetic bacteria in batch experiment. For controlling algal adhesion and enhancing the decolorant efficiency of photosynthetic bacteria, the possibility of using thin-film photocatalysis is implicated by the results. Carneiro et al. (2004) [129] studied the photobleaching of a textile azo dye and reactive orange 16 in aqueous solution using titanium dioxide thin-film electrodes. When 100% of color removal is obtained after 20 m of photo-electrocatalysis, it is recommended to have an applied potential of +1.0 V and low dye concentration. The final step of purification of pre-treated wastewaters utilizes the photocatalytic process because it is the most effective solutions with small amount of pollutants.

3.3.9 Enzymatic Treatment

Chemical reactions based on the action of biological catalysts are involved by the enzymatic system which falls between the two traditional categories of chemical and biological processes. Specifically, target components that are detrimental to the environment can be developed by an enzymatic process. The pre-treatment step to remove one or more compounds can interfere with subsequent downstream treatment processes, and enzymatic treatment can be used. The enzymatic treatment will be most effective in those streams that have the highest concentrations of target contaminants and the lowest concentration of other contaminants that may tend to interfere with enzymatic treatment in order to that of the susceptibility of enzymes to inactivation by the presence of the other chemicals. In white rot fungal cultures, it was examined that decolorization of eight synthetic dyes including azo, anthraquinone metal complex and indigo by peroxidase catalyzed oxidation. Manganese-dependent peroxidase (MnP) did not decolorize the dyes while above 80% color was removed by ligninase-catalyzed oxidation, and further dye decolorization rate is increased linearly with ligninase dosage (Lip) [130–132].

Chrysosporium were almost completely degraded some azo and heterocyclic dyes chrysosporium in ligninolytic solution but decolorized to a different extent (0–80%) by crude ligninase. Using chicken intestine proteolytic and autolytic enzymes, enzymatic hydrolysis of tannery flushings could remove 30% of the color.

The enzymatic process promotes quick decolorization of the dye; nevertheless, maximum decolorization degree of about 30% is insignificant in relation to the decolorization degree achieved by the other processes reported by the Zamera et al. [133]. For textile dye decolorization, the enzymatic activity of four white rot fungi is as follows: *P. tremellosa*, *P. ostreatus*, *B. adusta*, and *C. vesicolor*. The ability of white rot fungi is demonstrated in this study is to produce ligninolytic enzymes, which decolorize dyes in artificial effluents. For wastewater treatment, more specifically for dye decolorization, enzyme membrane reactors are emerging recently [134, 135].

3.3.10 Radiolysis

For the treatment of textile wastewater, ionizing radiation process is one of the promising methods because the effect of radiation can be intensified in aqueous solution, in which the dye molecules are degraded effectively by the primary products formed through radiolysis of water. Radiolysis process has sufficient energy to dislodge electrons from atoms and molecules used by the radiation and to convert them to electrically charged particles called ions. Further reactions of these species will lead usually reactive and that eventually will lead to chemical reactions in the formation of free radicals [136, 137].

3.3.11 Other Processes

The effect of photocatalytic and combined anaerobic-photocatalytic treatment of textile dyes mixed wastewaters were analyzed by Harrelkas et al. [138]. Photocatalytic process based on immobilized titanium dioxide was used for this study. On both autoxidized chemically reduced azo dyes and normal dyes, the treatment was applied. The photocatalysis process was able to remove more than 90% color from crude, and autoxidized chemically reduced dye solutions were found in the result [138, 139]. Removal of UV absorbance and COD was extended up to 50% in average. A biological process consisting an anaerobic process and aerobic process in series is a promising method for treating textile wastewater which has been reported by several scientific researchers. At the first stage, the anaerobic microorganisms reduced the color and increased the biodegradability of the wastewater, and in the second stage, the reduction in the bacterial count will render the effluent more amenable to subsequent aerobic treatment in this method [140].

- Combination of advanced oxidation processes,
- Different photochemical processes,
- Photochemical/electrochemical,
- Ultrasonic and other advanced oxidation processes,
- Coagulation-based combinations,

- Adsorption-based combinations,
- Membrane-based combinations,
- Biological treatment-based combinations,
- Combination among biological processes,
- Hybrid technologies based on biological processes.

3.4 Sustainable Sludge Management

The residual, semisolid material left from textile wastewater or other industrial wastewater is referred as sludge (Fig. 30). The chemical sludge tends to have concentrations of heavy metals due to the physical–chemical processes involved in the treatment. On the other hand, biological sludge is wealthy in nutrients such as nitrogen and phosphorous along with valuable organic matter that is beneficial to depleted soils or soils focus to erosion. An environmentally sensitive problem is the treatment and disposal of sludge. Sludge production will increase subsequently with the increase in a number of wastewater treatment plants and the evolution of more stringent environmental quality standards which is the mounting global problem. The major problem of water pollution control in terms of their immediate offensive nature and potential for pollution is regarded as sludge. Due to a few environmentally unfriendly disposal routes like land disposal and direct disposal to the sea having been phased out, finding cost-effective and advanced solutions while responding to environmental, regulatory, and public pressures is the challenge faced by the sludge management. For sustainable development, the recycling and use of wastes are the preferred options rather than incineration or land filling, because of sensitivities of such contaminants like heavy metals, but sometimes it is not straightforward. The disposal of sludge always requires careful management whether it is ease or difficulty, the disposal is actually achieved, and the associated

Fig. 30 Typical sludge formed from the textile effluent

costs depends very much on circumstances. Sludge management has considerable influence by the local and national geographical, agronomic, economic, and stakeholder perception factors [141, 142].

Methods of sludge treatment
Sludge is treated to change its chemical and/or physical properties to make it fit for environmentally safe disposal options before the disposal. The quantity and properties of chemical sludge varied and it depends on the process, therefore it is requied to do the stabilization process. Stabilization treatments will decrease the volume, a number of pathogenic organisms, concentrates contaminants and also diminishes odor is mainly determined by the wastewater composition, chemicals used, and treatment units. Hence, the transport of sludge is stabilized to make safe and cheaper. By conditioning and thickening, sludge can be treated to improve the effects of dewatering. On the sludge, it also has a disinfection and odor reduction effect. The four common conditioning techniques are gravity thickeners, gravity belt thickeners, dissolved air floatation, and drum thickeners, and the most common conditioning technique is chemical or thermal conditionings. Due to the cost component of chemicals, the chemical conditioning can also have a stabilizing effect as well as deodorizing. Thermal conditioning is suitable for all types of sludge. It requires energy, and there are also chances of a foul odor. To reduce water content and increase dry solid content, these methods are used. Ways to dewater the sludge have many ways in which some of them are as follows: the use of centrifuges, filter presses, recessed-plate filter presses, and a bed of reed or drying beds. Instead of dewatered, sludge can also be dried. To the sludge, the heat can be transferred either directly or indirectly through direct contact with the sludge or through a heat transfer surface. To condense and to ameliorate sludge which is produced by wastewater/effluent treatment, the biological sludge treatment is essential. Some of the treatment methods are [141–146]:

Anaerobic digestion: It is the decomposition of sludge in an anaerobic environment. The important features of anaerobic digestion are the mass reduction, methane production, and improved dewatering properties of the fermented sludge. However, for a digestion chamber, etc., it requires investment and it also has a slow degeneration rate. Pre-treatments such as thermal pre-treatment, the addition of enzymes, ozonation, chemical solubilization by acidification or alkaline hydrolysis, and mechanical sludge disintegration and ultrasound pre-treatment can be done to improve the biodegradability of sludge. By the high concentration of inhibitors such as heavy metals, temperature, etc., can also inhibit the degeneration process [147, 148].

Aerobic digestion: In the sludge, the process of oxidizing and decomposing the organic matter sludge by microorganisms in an aerobic environment is aerobic digestion. To reduce both the organic content and the volume of the sludge Aerobic sludge digestion is one process that may be used. Even though this process is sensitive to temperature, heavy metals, etc., and incurs higher energy costs, there is no useful by-product such as methane [149–151].

Enzyme treatment: To improve the hydrolysis of organic matter in sludge which limits the rate of anaerobic digestion resulting in an improvement in both biogas productions, this method is usually used as a pre-treatment of sludge and also improves the dewatering properties of the digested sludge [19].

Thermal Hydrolysis: To reduce sludge production in the sludge treatment process, thermal treatment is incorporated, anaerobic digesters and inactivate pathogens improve dewatering of sludge which enhance biogas production. In order to make sludge more readily biodegradable, this process hydrolyzes organic solids and makes them soluble. In the anaerobic digester, it also reduces the hydraulic retention time [152, 153].

Chemical stabilization: To stabilize the sludge solids, sludge is treated with chemicals in different ways in this process. Lime stabilization and chlorine stabilization are the common methods of chemical stabilization. All kinds of sludge eliminate odor and destroy pathogenic microorganisms which are stabilized by these methods. Due to their success of improving process yields, use of polyelectrolytes as a conditioner for sludge dewatering operations is also gaining popularity [19, 152, 154].

Microbial consortium: Sludge can be degraded by consortium of beneficial microbes. It can compile aerobic and/or anaerobic microorganisms. The microbial ecology of the sludge is modified by the application of microbial consortium. Maximum biological growth is achieved by the help of it, and as a result, the maximum organic matter will be consumed. BOD, COD, TSS, total and fecal coli form in all sewage and effluent generated in an industry are reduced by the microbial consortium like effective microorganism (EM). The putrefaction of sludge leading to odor reduction is suppressed by the consortium EM, and the microbes also secrete some enzymes which digest their own cell, thereby reducing the biomass/sludge. The need for use of extra chemical is also eliminated [146, 155, 156].

4 Case Studies

The case studies were conducted in the four different textile industries which are located in and around Tirupur, India. During the study, various pollutant parameters were analyzed, such as BOD, COD, TDS, TSS, other salts, minerals, color, and efficiency. The case study was conducted on daily basis for the period of 30 days to make average and tabulated (Tables from 15, 16, 17, 18 and 19), and since it is a critical issue to disclose the name of industries, it calls it as W–Z. The results of case study victimize that various pollutant parameters are reduced significantly and within their permissible limits which is suggested by central pollution control board (Fig. 31; Table 20).

Table 15 Performance of treatment system for wash water (data was collected from Textile plant "W", at Perundurai, TN, India)

Parameters	Wash water			RO-I		RO-II	
	Inlet	Primary treatment	Ozonation	Permeate	Reject	Permeate	Reject
pH	9.7	9.65	6.87	6.12	6.84	5.85	7.09
TSS (mg/l)	177	52	21	7	74	7	128
TDS (mg/l)	3150	1944	3325	944	9978	198	18,254
BOD (mg/l)	192	45	42	1	15	2	210
COD (mg/l)	598	172	132	25	326	19	744
Total hardness (mg/l)	95	NA	NA	NA	62	Nil	1075
Chlorides (mg/l)	345	640	688	321	109	36	4511
Color %	<10	<10	Colorless	Colorless	<10	Colorless	10–20

Table 16 Characteristics of effluents in Plant "X" at Tirupur, India

Parameters	Wash water		Dye bath water			
	Inlet to ETP	Outlet	RO permeate	RO reject	Dye bath wastewater	NF reject
pH	9.8	9.82	7.5	8.17	10.38	8.17
TSS (mg/l)	47	25	BDL	45	78	62
TDS (mg/l)	4300	3650	475	21,650	39,800	48,300
BOD (mg/l)	80	65	11	460	189	100
COD (mg/l)	321	209	25	1170	897	432
Total hardness (mg/l of $CaCO_3$)	344	145	3	758	98	415
Sulfate (mg/l)	85	125	7	330	175	385
Chlorine (mg/l)	1878	1698	197	10,800	20,147	26,855
Sodium (mg/l)	1750	NA	NA	9800	NA	22,500
Potassium (mg/l)	42	NA	NA	235	NA	71

Note BDL below detection limit; *NA* not analyzed

Table 17 Pollutant removal efficiency in common effluent treatment Plant "X" at Tirupur, India

Parameters	Pollutant removal efficiency (%)
BOD	88–98
COD	91–97
TDS	80–97
Chloride	75–97
Sodium	85–97
Nitrogen	80–92
Phosphorus	90–94

Table 18 Characteristics of effluents in Plant "Y" at Tirupur, India

Parameters	Wash water		RO stages	
	Inlet to ETP	Primary/ozonation/adsorption	Permeate	Reject
pH	9.6	9.5	7.52	8.1
TSS (mg/l)	53	21	3	49
TDS (mg/l)	3710	3655	412	20,544
BOD (mg/l)	72	55	7	295
COD (mg/l)	344	220	20	1045
Total hardness (mg/l)	192	139	0	690
Chlorides (mg/l)	1785	1777	184	10,214

Table 19 Characteristics of effluents in Plant "Z" at Tirupur, India

Parameters	Inlet effluent Inlet to ETP	Primary/ adsorption	Ultrafiltration	RO stages	
				Permeate	Reject
pH	9.24	9.01	6.98	7.12	7.44
TSS (mg/l)	165	76	35	11	87
TDS (mg/l)	11,477	8445	5144	654	12,875
BOD (mg/l)	235	175	6	1	8
COD (mg/l)	568	335	84	78	114
Total hardness (mg/l)	98	NA	NA	84	2740
Chlorides (mg/l)	4870	3325	NA	235	4787
Sulfate (mg/l)	1445	815	278	18	515
Color (Hazen)	260	55	20	20	25

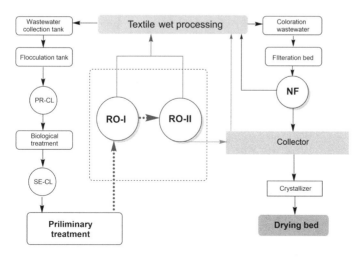

Fig. 31 Typical schematic diagram of wastewater treatment in Plant "X" at Tirupur, India (PR-CL: primary clarifier, SE-CL: secondary clarifier, RO: reverse osmosis, NF: nanofiltration)

Table 20 Characteristics of effluents in Plant "Z" at Tirupur, India, with different days

Day	Parameters	ETP feed	Primary clarifier outlet	Secondary clarifier outlet	RO-1 feed	Evaporator-I feed	Crystallizer feed	Evaporator-II feed
Day-1	pH	8.87	6.44	7.38	6.64	8.02	8.79	8.68
	BOD (mg/L)	170	130	18	10	NA	NA	NA
	COD (mg/L)	648	392	132	82	660	NA	NA
	TDS (mg/L)	7800	8000	7600	8000	55,000	297,000	224,400
	TSS (mg/L)	168	114	40	<1	NA	NA	NA
	TS (TDS + TSS) (mg/L)	7968	8114	7640	8000	NA	NA	NA
	Turbidity	77.2	48.3	4.16	0.2	0.55	NA	NA
Day-2	pH	9.46	6.50	7.46	6.95	7.90	8.77	9.03
	BOD (mg/L)	136	110	22	12	NA	NA	NA
	COD (mg/L)	584	330	115	69	640	NA	NA
	TDS (mg/L)	8900	8600	8000	7800	52,800	294,600	230,600
	TSS (mg/L)	198	124	36	<1	NA	NA	NA
	TS (TDS + TSS) (mg/L)	9098	8724	8036	7800	NA	NA	NA
	Turbidity	90.5	41.5	4.7	0.19	0.58	NA	NA
Day-3	pH	10.05	6.61	7.74	6.67	7.98	9.12	9.18
	BOD (mg/L)	142	106	30	14	NA	NA	NA
	COD (mg/L)	612	390	145	93	664	NA	NA
	TDS (mg/L)	8300	8000	7600	7500	54,200	298,900	228,000
	TSS (mg/L)	174	126	42	<1	NA	NA	NA
	TS (TDS + TSS) (mg/L)	8474	8126	7642	7500	NA	NA	NA
	Turbidity	84.7	55.7	5.7	0.24	0.74	NA	NA

(continued)

Table 20 (continued)

Day	Parameters	ETP feed	Primary clarifier outlet	Secondary clarifier outlet	RO-1 feed	Evaporator-I feed	Crystallizer feed	Evaporator-II feed
Day-4	pH	10.13	6.98	7.78	6.65	8.18	9.28	9.35
	BOD (mg/L)	136	102	26	12	NA	NA	NA
	COD (mg/L)	608	385	140	91	640	NA	NA
	TDS (mg/L)	9100	8900	8400	8500	53,800	303,600	233,400
	TSS (mg/L)	168	130	48	<1	NA	NA	NA
	TS (TDS + TSS) (mg/L)	9268	9030	8448	8500	NA	NA	NA
	Turbidity	77.6	46.4	6.23	0.24	NA	NA	NA
Day-5	pH	9.85	6.53	7.79	6.85	8.19	8.98	9.13
	BOD (mg/L)	160	108	38	20	NA	NA	NA
	COD (mg/L)	622	412	156	101	NA	NA	NA
	TDS (mg/L)	7800	8200	8200	8400	54,800	301,800	212,200
	TSS (mg/L)	178	102	40	<1	NA	NA	NA
	TS (TDS + TSS) (mg/L)	7978	8302	8240	8400	NA	NA	NA
	Turbidity	75	39.2	4.81	0.22	NA	NA	NA
Day-6	pH	10.18	6.80	7.99	6.88	8.15	8.91	9.05
	BOD (mg/L)	150	136	32	12	NA	NA	NA
	COD (mg/L)	610	390	140	92	NA	NA	NA
	TDS (mg/L)	8800	9000	9000	8800	55,800	295,200	225,400
	TSS (mg/L)	148	86	34	<1	NA	NA	NA
	TS (TDS + TSS) (mg/L)	8948	9086	9034	8800	NA	NA	NA
	Turbidity	68.2	34.3	4.05	0.26	NA	NA	NA

(continued)

74 A. P. Periyasamy et al.

Table 20 (continued)

Day	Parameters	ETP feed	Primary clarifier outlet	Secondary clarifier outlet	RO-1 feed	Evaporator-I feed	Crystallizer feed	Evaporator-II feed
Day-7	pH	9.88	6.85	7.98	6.94	8.15	9.38	9.45
	BOD (mg/L)	130	100	30	18	NA	NA	NA
	COD (mg/L)	544	340	124	78	668	NA	NA
	TDS (mg/L)	8400	8600	8400	8600	54,800	296,800	216,800
	TSS (mg/L)	270	148	38	<1	NA	NA	NA
	TS (TDS + TSS) (mg/L)	8670	8748	8438	8600	NA	NA	NA
	Turbidity	123	56.4	4.02	0.36	0.68	NA	NA
Day-8	pH	10.00	6.89	7.88	6.82	8.27	9.23	9.39
	BOD (mg/L)	110	96	24	10	NA	NA	NA
	COD (mg/L)	510	305	108	73	NA	NA	NA
	TDS (mg/L)	8400	8500	8400	8400	56,400	294,300	219,600
	TSS (mg/L)	288	136	60	<1	NA	NA	NA
	TS (TDS + TSS) (mg/L)	8688	8636	8460	8400	NA	NA	NA
	Turbidity	130	40.4	8.43	0.34	0.63	NA	NA
Day-9	pH	9.99	7.02	7.85	6.98	8.05	9.15	9.25
	BOD (mg/L)	120	90	20	10	NA	NA	NA
	COD (mg/L)	496	304	115	77	610	NA	NA
	TDS (mg/L)	8900	9100	8500	8700	53,400	293,600	220,600
	TSS (mg/L)	248	112	30	<1	NA	NA	NA
	TS (TDS + TSS) (mg/L)	9148	9212	8530	8700	NA	NA	NA
	Turbidity	103	40.1	3.99	0.26	0.4	NA	NA

(continued)

Table 20 (continued)

Day	Parameters	ETP feed	Primary clarifier outlet	Secondary clarifier outlet	RO-1 feed	Evaporator-I feed	Crystallizer feed	Evaporator-II feed
Day-10	pH	10.18	6.52	7.85	6.61	8.15	9.15	9.21
	BOD (mg/L)	102	80	10	Nil	NA	NA	NA
	COD (mg/L)	480	280	86	53	540	NA	NA
	TDS (mg/L)	7400	7600	7500	8100	54,200	296,800	219,200
	TSS (mg/L)	184	98	42	<1	NA	NA	NA
	TS (TDS + TSS) (mg/L)	7584	7698	7542	8100	NA	NA	NA
	Turbidity	86	39.3	6.98	0.28	0.66	NA	NA
Day-11	pH	9.59	6.62	7.61	6.59	7.86	9.11	9.15
	BOD (mg/L)	120	70	26	12	NA	NA	NA
	COD (mg/L)	568	352	130	84	NA	NA	NA
	TDS (mg/L)	7000	7400	7000	7600	54,700	297,100	227,300
	TSS (mg/L)	176	78	48	<1	NA	NA	NA
	TS (TDS + TSS) (mg/L)	7176	7478	7048	7600	NA	NA	NA
	Turbidity	86.8	26.4	6.48	0.31	NA	NA	NA
Day-12	pH	8.98	6.38	7.78	6.51	7.79	9.07	9.21
	BOD (mg/L)	104	80	22	12	NA	NA	NA
	COD (mg/L)	502	315	105	68	NA	NA	NA
	TDS (mg/L)	7200	7400	7600	7800	55,300	299,200	226,100
	TSS (mg/L)	170	96	47	<1	NA	NA	NA
	TS (TDS + TSS) (mg/L)	7370	7496	7664	7800	NA	NA	NA
	Turbidity	76.4	37.3	9.6	0.29	NA	NA	NA

(continued)

Table 20 (continued)

Day	Parameters	ETP feed	Primary clarifier outlet	Secondary clarifier outlet	RO-1 feed	Evaporator-I feed	Crystallizer feed	Evaporator-II feed
Day-13	pH	9.45	6.81	7.66	6.58	8.14	8.54	8.58
	BOD (mg/L)	160	114	44	22	NA	NA	NA
	COD (mg/L)	696	394	139	92	NA	NA	NA
	TDS (mg/L)	9800	9200	8600	8300	56,400	298,800	216,400
	TSS (mg/L)	238	98	50	<1	NA	NA	NA
	TS (TDS + TSS) (mg/L)	10,038	9298	8650	8300	NA	NA	NA
	Turbidity	103	39.4	7.15	0.31	NA	NA	NA
Day-14	pH	8.93	6.52	7.51	6.48	7.75	8.64	8.48
	BOD (mg/L)	Under test	Under test	Under test	Under test	NA	NA	NA
	COD (mg/L)	728	424	144	92	NA	NA	NA
	TDS (mg/L)	8800	9400	8600	8200	54,300	299,800	218,600
	TSS (mg/L)	234	104	46	<1	NA	NA	NA
	TS (TDS + TSS) (mg/L)	9034	9504	8646	8200	NA	NA	NA
	Turbidity	110	39.8	6.31	0.31	NA	NA	NA
Day-15	pH	8.39	6.23	7.79	6.95	7.85	8.95	8.82
	BOD (mg/L)	Under test	Under test	Under test	Under test	NA	NA	NA
	COD (mg/L)	720	435	156	101	NA	NA	NA
	TDS (mg/L)	8600	9000	9100	8900	53,200	298,600	214,600
	TSS (mg/L)	238	118	47	<1	NA	NA	NA
		8838	9118	9147	8900	NA	NA	NA

(continued)

Table 20 (continued)

Day	Parameters	ETP feed	Primary clarifier outlet	Secondary clarifier outlet	RO-1 feed	Evaporator-I feed	Crystallizer feed	Evaporator-II feed
	TS (TDS + TSS) (mg/L)							
	Turbidity	110	41.4	9.1	0.31	NA	NA	NA
Day-16	pH	9.34	6.67	7.78	6.48	8.11	9.32	9.18
	BOD (mg/L)	Under test	Under test	Under test	Under test	NA	NA	NA
	COD (mg/L)	560	355	128	81	NA	NA	NA
	TDS (mg/L)	9100	9000	8800	8800	54,600	298,600	216,200
	TSS (mg/L)	148	102	51	<1	NA	NA	NA
	TS (TDS + TSS) (mg/L)	9248	9102	8851	8800	NA	NA	NA
	Turbidity	64.2	29	11	0.35	NA	NA	NA

5 Conclusions

This chapter summarized the various types of sustainable wastewater treatments for textile industry, and also it described the environmental hazards associated with conventional wastewater treatment techniques. This chapter clearly depicted the different characteristics, such as BOD, COD, TDS, TSS, turbidity, and pH variation of pollution caused by the each stage of textile processing with respect to utilization various chemicals and auxiliaries. The various sustainable wastewater treatment techniques like AC adsorption, AOP, electrocoagulant, membrane technology, ultrasonic associated effluent treatments, photocatalysis, SBR/MBR are reviewed deeply with their pros and cons as well as their pollutant removal efficiency. Generally, the AC adsorption and ozonation treatments are used to prepare the effluent for membrane filtration technique to get adequate process efficiency. Use of multiple effect evaporation system could play vital role to reduce the effluent volume. The multiple effect evaporation techniques can help to recover the glauber salt and other residual substances. Recently, the reuse of effluent treated water is assumed significant importance, due to scarcity of water resources and increasingly stringent regulatory requirements for the disposal of the effluent. However, reuse or recycle of wastewater depends on the selection of effluent treatment process.

5.1 Recommendations for Effluent Management in Textile Industries

Textile industry treated the effluent to segregate the water and other substance; generally the recovery of water is quite easy than the salt or other residues. It is better to treat the highly polluting effluents such as dye bath discharge; however, it is 10% of total effluent discharge, and remaining 90% are from low-polluting streams like wash water. Segregation of these streams makes huge difference in the effluent treatment, since low-polluting streams have nominal dissolved solids, heavy metals, and other pollutant materials. Sustainable wastewater treatment can be appropriately applied on both the effluent streams; however, the sustainable wastewater treatments have the scope to recover and recycle the water and other substances. For the effective effluent treatment for textile industries, there are some recommendations,

- High pollutant effluent stream can be treated separately, when it mixed with low pollutant streams, it makes difficult in the treatment and it increases the overall cost.
- The reject of EC, RO, NF should be treated with high pollutant effluent stream.
- Utilize maximum to sustainable wastewater treatment, since it produces less residue (sludge).

- Sustainable wastewater treatment has no compliance with respect to the norms and standards.
- Combination of two or more processes (i.e., EC/Membrane filtration) makes better pollutant removal efficiency.

References

1. ONU (2017) World population to hit 9.8 billion by 2050, despite nearly universal lower fertility rates—UN. 2015–2017
2. Muthu SS (2014) Ways of measuring the environmental impact of textile processing: an overview. In: Assessing environment impact textile clothing supply chain. Elsevier, pp 32–56
3. Karthik T, Gopalakrishnan D (2015) Roadmap to sustainable textiles and clothing. In: Muthu SS (ed) Roadmap to sustainable textile clothing environment social aspect textile clothing supply chain. Springer, Singapore, pp 153–188
4. Chequer FMD, de Oliveira GAR, Ferraz ERA et al (2013) Textile dyes: dyeing process and environmental impact. In: Günay M (ed) Eco-friendly textile dyeing. Finish. pp 151–176
5. Hauser PJ (2000) Reducing pollution and energy requirements in cotton dyeing. Text Chem Color Am Dyest Rep 32:44–48
6. Periyasamy AP, Duraisamy G (2018) Carbon footprint on denim manufacturing. In: Martínez LMT, Kharissova OV, Kharisov BI (eds) Handbook of Ecomaterials. Springer International Publishing, Cham, pp 1–18
7. Periyasamy AP, Dhurai B, Thangamani K (2011) Salt-free dyeing—a new method of dyeing on lyocell/cotton blended fabrics with reactive dyes. Autex Res J 11:14–17. http://www.autexrj.com/cms/zalaczone_pliki/3_01_11.pdf
8. Periyasamy AP, Venkatesan H (2018) Eco-materials in textile finishing. In: Martínez LMT, Kharissova OV, Kharisov BI (eds) Handbook of Ecomaterials. Springer International Publishing, Cham, pp 1–22
9. Periyasamy AP, Vikova M, Vik M (2017) A review of photochromism in textiles and its measurement. Text Prog 49:53–136. http://doi.org/10.1080/00405167.2017.1305833
10. Periyasamy AP (2016) Effect of PVAmHCl pre-treatment on the properties of modal fabric dyed with reactive dyes: an approach for salt free dyeing. J Text Sci Eng 6(262):1–9. http://doi.org/10.4172/2165-8064.1000262
11. US EPA O (2016) Effluent guidelines. Environ Prot Agency 15–19
12. Central Pollution control board (2010) Standards for effluents from textile industry. Environ Rules 1986 1–16. http://cpcb.nic.in/industry-effluent-standards/
13. Shabbir M, Mohammad F (2017) Sustainable production of regenerated cellulosic fibres. In: Muthu SS (eds) Sustainable fibres textiles. Woodhead Publishing, Sawston, pp 171–189
14. Chen J (2014) Synthetic textile fibers: regenerated cellulose fibers. In: Sinclair R (ed) Textile fashion materials design technology. Woodhead Publishing, Sawston, pp 79–95
15. El-Halwagi MM (1997) Synthesis of reactive mass-exchange networks, pollution prevention through process integration. Academic Press, San Diego, pp 191–216
16. Abdel-Halim ES (2012) Simple and economic bleaching process for cotton fabric. Carbohydr Polym 88:1233–1238. https://doi.org/10.1016/j.carbpol.2012.01.082
17. Imran MA, Hussain T, Memon MH, Abdul Rehman MM (2015) Sustainable and economical one-step desizing, scouring and bleaching method for industrial scale pretreatment of woven fabrics. J Clean Prod 108:494–502. https://doi.org/10.1016/j.jclepro.2015.08.073

18. Khatri A, Peerzada MH, Mohsin M, White M (2015) A review on developments in dyeing cotton fabrics with reactive dyes for reducing effluent pollution. J Clean Prod 87:50–57. https://doi.org/10.1016/j.jclepro.2014.09.017

19. Vigneswaran C, Ananthasubramanian M, Kandhavadivu P et al (2014) 5—enzymes in textile effluents. In: Vigneswaran C, Ananthasubramanian M, Kandhavadivu P (eds) Bioprocessing textiles. Woodhead Publishing, New Delhi, pp 251–298

20. Khatri A, White M (2015) Sustainable dyeing technologies. In: Blackburn R (ed) Sustainable apparel production processing recycling. Woodhead Publishing, Sawston, pp 135–160

21. Kumar PS, Narayan AS, Dutta A (2017) Textiles and clothing sustainability. In: Muthu SS (ed) Textile clothing sustainable textile science clothing technology. Springer, Singapore, pp 57–96

22. (1998) Dye and pigment manufacturing industry—pollution prevention guidelines. https://www.environmental-expert.com/articles/dye-and-pigment-manufacturing-industry-pollution-prevention-guidelines-1380. Accessed 6 Dec 2017

23. Periyasamy AP, Ramamoorthy SK, Lavate SS (2018) Eco-friendly Denim Processing. In: Martínez LMT, Kharissova OV, Kharisov BI (eds) Handbook of Ecomaterials. Springer International Publishing, Cham, pp 1–21

24. Periyasamy AP, Rwahwire S, Zhao Y (2018) Environmental Friendly Textile Processing. In: Martínez LMT, Kharissova OV, Kharisov BI (eds) Handbook of Ecomaterials. Springer International Publishing, Cham, pp 1–38

25. Venkatesan H, Periyasamy AP (2017) Eco-fibers in the textile industry. In: Martínez LMT, Kharissova OV, Kharisov BI (eds) Handbook of Ecomaterials. Springer International Publishing, Cham, pp 1–21

26. Periyasamy AP, Militky J (2017) Denim and consumers' phase of life cycle. In: Muthu SS (ed) Sustainability Denim. Woodhead Publishing Limited, U.K, pp 257–282

27. Periyasamy AP, Wiener J, Militky J (2017) Life-cycle assessment of denim. In: Muthu SS (ed) Sustainability Denim. Woodhead Publishing Limited, U.K, pp 83–110

28. Periyasamy AP, Militky J (2017) Denim processing and health hazards. In: Muthu SS (ed) Sustainability Denim. Woodhead Publishing Limited, U.K, pp 161–196

29. Viková M, Periyasamy AP, Vik M, Ujhelyiová A (2017) Effect of drawing ratio on difference in optical density and mechanical properties of mass colored photochromic polypropylene filaments. J Text Inst 108:1365–1370. http://doi.org/10.1080/00405000.2016.1251290

30. Gähr F, Hermanutz F, Oppermann W (1994) Ozonation- an important technique to comly with new German laws for textile waste-water treatment. Water Sci Technol 30:255–263

31. Aljeboree AM, Alshirifi AN, Alkaim AF (2017) Kinetics and equilibrium study for the adsorption of textile dyes on coconut shell activated carbon. Arab J Chem 10:S3381–S3393. https://doi.org/10.1016/j.arabjc.2014.01.020

32. Wang W, Deng S, Li D et al (2018) Sorption behavior and mechanism of organophosphate flame retardants on activated carbons. Chem Eng J 332:286–292. https://doi.org/10.1016/j.cej.2017.09.085

33. Kismir Y, Aroguz AZ (2011) Adsorption characteristics of the hazardous dye brilliant green on Sakli{dotless}kent mud. Chem Eng J 172:199–206. https://doi.org/10.1016/j.cej.2011.05.090

34. Khaled A, El Nemr A, El-Sikaily A, Abdelwahab O (2009) Treatment of artificial textile dye effluent containing direct yellow 12 by orange peel carbon. Desalination 238:210–232. https://doi.org/10.1016/j.desal.2008.02.014

35. Koltuniewicz AB (2017) Process intensification: definition and application to membrane processes. In: Figoli A, Criscuoli A (eds) Sustainable membrane technology water wastewater treatment. Springer, Singapore, pp 67–96

36. Cho J, Ramírez JAL, Shon HK et al (2012) Sustainable water treatment sustainability/sustainable water treatment using nanofiltration and tight ultrafiltration membranes. In:

Meyers RA (ed) Encyclopedia sustainability science technology. Springer, New York, pp 10530–10542

37. Mauskar JM (2007) Advance methods for treatment of textile industry effleunts. Cent Pollut Control Board-India 1–137. http://cpcb.nic.in/newitems/27.pdf

38. Mamigonyan RA, Gutin YV (2003) Creation of a new generation of micro- and ultra-filtration units for separation of aggressive waste water. Chem Pet Eng 39:442–445. https://doi.org/10.1023/A:1026301332719

39. Metz S, Gawad S, Trautmann C et al (2002) Polyimide-based microfluidic devices with nanoporous membranes for filtration and separation of particles and molecules. In: Baba Y, Shoji S, van den Berg A (eds) Micro total analysis systems 2002. Springer, Dordrecht, pp 727–729

40. Bhave RR, Guibaud J, Rumeau R (1991) Inorganic membranes for the filtration of water, wastewater treatment and process industry filtration applications. In: Bhave RR (ed) Inorganic membrane synthesis characteristics application. Springer, Dordrecht, pp 275–299

41. Cséfalvay E, Imre P, Mizsey P (2008) Applicability of nanofiltration and reverse osmosis for the treatment of wastewater of different origin. Open Chem 6:277–283. https://doi.org/10.2478/s11532-008-0026-3

42. Delyannis A, Delyannis E-E (1980) Other applications of reverse osmosis and ultrafiltration. In: Delyannis A, Delyannis E-E (eds) Seawater desalt. Springer, Berlin, pp 160–168

43. Singh KP, Mohan D, Sinha S et al (2003) Color removal from wastewater using low-cost activated carbon derived from agricultural waste material. Ind Eng Chem Res 42:1965–1976. https://doi.org/10.1021/ie020800d

44. Malik PK (2004) Dye removal from wastewater using activated carbon developed from sawdust: adsorption equilibrium and kinetics. J Hazard Mater 113:81–88. https://doi.org/10.1016/j.jhazmat.2004.05.022

45. Sulaymon AH, Abood WM (2014) Removal of reactive yellow dye by adsorption onto activated carbon using simulated wastewater. Desalin Water Treat 52:3421–3431. https://doi.org/10.1080/19443994.2013.800341

46. Qiu M, Huang C (2015) Removal of dyes from aqueous solution by activated carbon from sewage sludge of the municipal wastewater treatment plant. Desalin Water Treat 53:3641–3648. https://doi.org/10.1080/19443994.2013.873351

47. Acharya J, Sahu JN, Mohanty CR, Meikap BC (2009) Removal of lead (II) from wastewater by activated carbon developed from <i> Tamarind wood </i> by zinc chloride activation. Chem Eng J 149:249–262. https://doi.org/10.1016/j.cej.2008.11.035

48. Tsai WT, Chang CY, Lin MC et al (2001) Characterization of activated carbons prepared from sugarcane bagasse by zncl 2 activation. J Environ Sci Heal Part B 36:365–378. https://doi.org/10.1081/PFC-100103576

49. Namasivayam C, Kavitha D (2002) Removal of Congo Red from water by adsorption onto activated carbon prepared from coir pith, an agricultural solid waste. Dye Pigm 54:47–58. https://doi.org/10.1016/S0143-7208(02)00025-6

50. Ahmad MA, Rahman NK (2011) Equilibrium, kinetics and thermodynamic of Remazol Brilliant Orange 3R dye adsorption on coffee husk-based activated carbon. Chem Eng J 170:154–161. https://doi.org/10.1016/j.cej.2011.03.045

51. Özhan A, Şahin Ö, Küçük MM, Saka C (2014) Preparation and characterization of activated carbon from pine cone by microwave-induced $ZnCl_2$ activation and its effects on the adsorption of methylene blue. Cellulose 21:2457–2467. https://doi.org/10.1007/s10570-014-0299-y

52. Moreno-Piraján JC, Garcia-Cuello VS, Giraldo L (2011) The removal and kinetic study of Mn, Fe, Ni and Cu ions from wastewater onto activated carbon from coconut shells. Adsorption 17:505–514. https://doi.org/10.1007/s10450-010-9311-5

53. Senthilkumaar S, Kalaamani P, Subburaam CV (2006) Liquid phase adsorption of crystal violet onto activated carbons derived from male flowers of coconut tree. J Hazard Mater 136:800–808. https://doi.org/10.1016/j.jhazmat.2006.01.045

54. Astashkina OV, Bogdan NF, Lysenko AA, Kuvaeva EP (2008) Production of activated carbon fibres by solid-phase (chemical) activation. Fibre Chem 40:179–185. https://doi.org/10.1007/s10692-008-9034-5

55. Osintsev KV, Osintsev VV, Dzhundubaev AK et al (2013) The production of activated carbon using the equipment of thermal power plants and heating plants. Therm Eng 60:583–590. https://doi.org/10.1134/S0040601513070082

56. Zhang ZJ, Cui P, Chen XY, Liu JW (2013) The production of activated carbon from cation exchange resin for high-performance supercapacitor. J Solid State Electrochem 17:1749–1758. https://doi.org/10.1007/s10008-013-2039-x

57. Tang S, Yuan D, Zhang Q et al (2016) Fe-Mn bi-metallic oxides loaded on granular activated carbon to enhance dye removal by catalytic ozonation. Environ Sci Pollut Res 23:18800–18808. https://doi.org/10.1007/s11356-016-7030-5

58. Hung Y-T, Lo HH, Wang LK et al (2005) Granular activated carbon adsorption. In: Wang LK, Hung Y-T, Shammas NK (eds) Physicochemical treatment process. Humana Press, Totowa, pp 573–633

59. Lourenço ND, Franca RDG, Moreira MA et al (2015) Comparing aerobic granular sludge and flocculent sequencing batch reactor technologies for textile wastewater treatment. Biochem Eng J 104:57–63. https://doi.org/10.1016/j.bej.2015.04.025

60. Belaid KD, Kacha S, Kameche M, Derriche Z (2013) Adsorption kinetics of some textile dyes onto granular activated carbon. J Environ Chem Eng 1:496–503. https://doi.org/10.1016/j.jece.2013.05.003

61. Zeng Q, Hao T, Mackey HR et al (2017) Alkaline textile wastewater biotreatment: a sulfate-reducing granular sludge based lab-scale study. J Hazard Mater 332:104–111. https://doi.org/10.1016/j.jhazmat.2017.03.005

62. Peláez-Cid A-A, Herrera-González A-M, Salazar-Villanueva M, Bautista-Hernández A (2016) Elimination of textile dyes using activated carbons prepared from vegetable residues and their characterization. J Environ Manage 181:269–278. https://doi.org/10.1016/j.jenvman.2016.06.026

63. Abou-Elela SI, Ali MEM, Ibrahim HS (2016) Combined treatment of retting flax wastewater using Fenton oxidation and granular activated carbon. Arab J Chem 9:511–517. https://doi.org/10.1016/j.arabjc.2014.01.010

64. Wang LK, Shammas NK, Williford C et al (2006) Evaporation processes. In: Wang LK, Shammas NK, Hung Y-T (eds) Advanced physicochemical treatment process. Humana Press, Totowa, pp 549–579

65. Pletcher D, Walsh FC (1993) Water purification, effluent treatment and recycling of industrial process streams. In: Pletcher D, Walsh FC (eds) Industrial electrochemistry. Springer, Dordrecht, pp 331–384

66. Srithar K, Mani A (2006) Studies on solar flat plate collector evaporation systems for tannery effluent (soak liquor). J Zhejiang Univ A 7:1870–1877. https://doi.org/10.1631/jzus.2006.A1870

67. Sarayu K, Sandhya S (2012) Current technologies for biological treatment of textile wastewater-a review. Appl Biochem Biotechnol 167:645–661. https://doi.org/10.1007/s12010-012-9716-6

68. Ranganathan K, Karunagaran K, Sharma DC (2007) Recycling of wastewaters of textile dyeing industries using advanced treatment technology and cost analysis—case studies. Resour Conserv Recycl 50:306–318. https://doi.org/10.1016/j.resconrec.2006.06.004

69. Vik EA, Carlson DA, Eikum AS, Gjessing ET (1984) Electrocoagulation of potable water. Water Res 18:1355–1360. https://doi.org/10.1016/0043-1354(84)90003-4

70. Matteson MJ, Dobson RL, Glenn RW et al (1995) Electrocoagulation and separation of aqueous suspensions of ultrafine particles. Colloids Surf A Physicochem Eng Asp 104:101–109. https://doi.org/10.1016/0927-7757(95)03259-G

71. Kabdaşlı I, Arslan-Alaton I, Ölmez-Hancı T, Tünay O (2012) Electrocoagulation applications for industrial wastewaters: a critical review. Environ Technol Rev 1:2–45. https://doi.org/10.1080/21622515.2012.715390

72. Barrera-Díaz C, Bilyeu B, Roa G, Bernal-Martinez L (2011) Physicochemical aspects of electrocoagulation. Sep Purif Rev 40:1–24. https://doi.org/10.1080/15422119.2011.542737

73. Liu H, Zhao X, Qu J (2010) Electrocoagulation in water treatment. In: Comninellis C, Chen G (eds) Electrochemistry environmental. Springer, New York, pp 245–262

74. Van der Bruggen B, Canbolat ÇB, Lin J, Luis P (2017) The potential of membrane technology for treatment of textile wastewater. In: Figoli A, Criscuoli A (eds) Sustainable membrane technology water wastewater treatment. Springer, Singapore, pp 349–380

75. Sahu O, Mazumdar B, Chaudhari PK (2014) Treatment of wastewater by electrocoagulation: a review. Environ Sci Pollut Res 21:2397–2413. https://doi.org/10.1007/s11356-013-2208-6

76. Islam SMD-U (2017) Electrocoagulation (EC) technology for wastewater treatment and pollutants removal. Sustain Water Resour Manag. https://doi.org/10.1007/s40899-017-0152-1

77. Yuksel E, Eyvaz M, Gurbulak E (2013) Electrochemical treatment of colour index reactive orange 84 and textile wastewater by using stainless steel and iron electrodes. Environ Prog Sustain Energy 32:60–68. https://doi.org/10.1002/ep.10601

78. Khandegar V, Saroha AK (2014) Electrochemical treatment of textile effluent containing acid red 131 dye. J Hazard Toxic Radioact Waste 18:38–44. https://doi.org/10.1061/(ASCE) HZ.2153-5515.0000194

79. Patel UD, Ruparelia JP, Patel MU (2011) Electrocoagulation treatment of simulated floor-wash containing reactive Black 5 using iron sacrificial anode. J Hazard Mater 197:128–136. https://doi.org/10.1016/j.jhazmat.2011.09.064

80. Can OT, Bayramoglu M, Kobya M (2003) Decolorization of reactive dye solutions by electrocoagulation using aluminum electrodes. Ind Eng Chem Res 42:3391–3396. https://doi.org/10.1021/ie020951g

81. Phalakornkule C, Polgumhang S, Tongdaung W (2009) Performance of an electrocoagulation process in treating direct dye: batch and continuous upflow processes. World Acad Sci Eng Technol 3:267–272

82. Daneshvar N, Sorkhabi HA, Kasiri M (2004) De-colourisation of dye containing acid red 14 by electrocoagulation with comparartive investigation of different electrode connections. J Hazard Mater 112:55–62. doi:https://doi.org/10.1016/j.jhazmat.2004.03.021

83. Fernandes A, Morão A, Magrinho M, Lopes A, Goncalves I (2004) Electrochemical degradation of C. I. Acid orange 7. Dye Pigm 61:287–296. doi:https://doi.org/10.1016/j.dyepig.2003.11.008

84. Merzouk B, Gourich B, Madani K et al (2011) Removal of a disperse red dye from synthetic wastewater by chemical coagulation and continuous electrocoagulation. A comparative study. Desalination 272:246–253. https://doi.org/10.1016/j.desal.2011.01.029

85. Mollah MYA, Morkovsky P, Gomes JAG et al (2004) Present and future perspectives of electrocoagulation. J Hazard Mater 114:199–210. https://doi.org/10.1016/j.jhazmat.2004.08.009

86. Khandegar V, Saroha AK (2013) Electrocoagulation for the treatment of textile industry effluent—a review. J Environ Manage 128:949–963. https://doi.org/10.1016/j.jenvman.2013.06.043

87. Bayramoglu M, Kobya M, Can OT, Sozbir M (2004) Operating cost analysis of electroagulation of textile dye wastewater. Sep Purif Technol 37:117–125. https://doi.org/10.1016/j.seppur.2003.09.002

88. Fajardo AS, Martins RC, Silva DR et al (2017) Dye wastewaters treatment using batch and recirculation flow electrocoagulation systems. J Electroanal Chem 801:30–37. https://doi.org/10.1016/j.jelechem.2017.07.015

89. Nasrullah M, Singh L, Krishnan S et al (2017) Electrode design for electrochemical cell to treat palm oil mill effluent by electrocoagulation process. Environ Technol Innov. https://doi.org/10.1016/j.eti.2017.10.001

90. Zakaria Z, Naje AS (2016) Electrocoagulation by rotated anode: a novel reactor design for textile wastewater treatment. J Environ Manage 176:34–44. https://doi.org/10.1016/j.jenvman.2016.03.034

91. Kuroda Y, Kawada Y, Takahashi T et al (2003) Effect of electrode shape on discharge current and performance with barrier discharge type electrostatic precipitaor. J Electrostat 57:407–415. https://doi.org/10.1016/S0304-3886(02)00177-8
92. Nielsen NF, Andersson C (2009) Electrode shape and collector plate spacing effects on ESP performance. In: Yan K (ed) Electrostatic precipitator. Springer, Berlin, pp 111–118
93. Mansoorian HJ, Mahvi AH, Jafari AJ (2014) Removal of lead and zinc from battery industry wastewater using electrocoagulation process: influence of direct and alternating current by using iron and stainless steel rod electrodes. Sep Purif Technol 135:165–175. https://doi.org/10.1016/j.seppur.2014.08.012
94. Eyvaz M, Kirlaroglu M, Aktas TS, Yuksel E (2009) The effects of alternating current electrocoagulation on dye removal from aqueous solutions. Chem Eng J 153:16–22. https://doi.org/10.1016/j.cej.2009.05.028
95. Vasudevan S, Lakshmi J (2011) Effects of alternating and direct current in electrocoagulation process on the removal of cadmium from water—a novel approach. Sep Purif Technol 80:643–651. https://doi.org/10.1016/j.jhazmat.2011.04.081
96. Verma SK, Khandegar V, Saroha AK (2013) Removal of chromium from electroplating industry effluent using electrocoagulation. J Hazard Toxic Radioact Waste 17:146–152. https://doi.org/10.1061/(ASCE)HZ.2153-5515.0000170
97. Modirshahla N, Behnajady MA, Mohammadi-Aghdam S (2008) Investigation of the effect of different electrodes and their connections on the removal efficiency of 4-nitrophenol from aqueous solution by electrocoagulation. J Hazard Mater 154:778–786. https://doi.org/10.1016/j.jhazmat.2007.10.120
98. Aoudj S, Khelifa A, Drouiche N et al (2010) Electrocoagulation process applied to wastewater containing dyes from textile industry. Chem Eng Process Process Intensif 49:1176–1182. https://doi.org/10.1016/j.cep.2010.08.019
99. Kobya M, Demirbas E, Can OT, Bayramoglu M (2006) Treatment of levafix orange textile dye solution by electrocoagulation. J Hazard Mater 132:183–188. https://doi.org/10.1016/j.jhazmat.2005.07.084
100. Rege MA, Bhojani SH, Tock RW, Narayan RS (1991) Advanced oxidation processes for destruction of dissolved organics in process wastewater: statistical design of experiments. Ind Eng Chem Res 30:2583–2586. https://doi.org/10.1021/ie00060a012
101. Deng Y, Zhao R (2015) Advanced oxidation processes (AOPs) in wastewater treatment. Curr Pollut Rep 1:167–176. https://doi.org/10.1007/s40726-015-0015-z
102. Snider EH, Porter JJ (1974) Ozone treatment of dye waste. J Water Pollut Control Fed 46:886–894
103. Demirev A, Nenov V (2005) Ozonation of two acidic azo dyes with different substituents. Ozone Sci Eng 27:475–485. https://doi.org/10.1080/01919510500351834
104. Selcuk H (2005) Decolorization and detoxification of textile wastewater by ozonation and coagulation processes. Dye Pigment 64:217–222. https://doi.org/10.1016/j.dyepig.2004.03.020
105. Liakou S, Kornaros M, Lyberatos G (1997) Pretreatment of azo dyes using ozone. Water Sci Technol 36:155–163. https://doi.org/10.1016/S0273-1223(97)00383-1
106. Lawrence J, Cappelli FP (1977) Ozone in drinking water treatment: a review. Sci Total Environ 7:99–108. https://doi.org/10.1016/0048-9697(77)90001-8
107. Fanchiang JM, Tseng DH (2009) Degradation of anthraquinone dye C.I. Reactive Blue 19 in aqueous solution by ozonation. Chemosphere 77:214–221. https://doi.org/10.1016/j.chemosphere.2009.07.038
108. Fu Z, Zhang Y, Wang X (2011) Textiles wastewater treatment using anoxic filter bed and biological wriggle bed-ozone biological aerated filter. Bioresour Technol 102:3748–3753. https://doi.org/10.1016/j.biortech.2010.12.002
109. Sarasa J, Roche MP, Ormad MP et al (1998) Treatment of a wastewater resulting from dyes manufacturing with ozone and chemical coagulation. Water Res 32:2721–2727. https://doi.org/10.1016/S0043-1354(98)00030-X

110. Zaror C, Carrasco V, Perez L et al (2001) Kinetics and toxicity of direct reaction between ozone and 1,2-dihydrobenzene in dilute aqueous solution. Water Sci Technol 43:321–326

111. Sevimli MF, Sarikaya HZ (2002) Ozone treatment of textile effluents and dyes: effect of applied ozone dose, pH and dye concentration. J Chem Technol Biotechnol 77:842–850. https://doi.org/10.1002/jctb.644

112. Paździor K, Wrębiak J, Klepacz-Smółka A et al (2017) Influence of ozonation and biodegradation on toxicity of industrial textile wastewater. J Environ Manage 195:166–173. https://doi.org/10.1016/j.jenvman.2016.06.055

113. Gharbani P, Tabatabaii SM, Mehrizad A (2008) Removal of Congo red from textile wastewater by ozonation. Int J Environ Sci Technol 5:495–500. https://doi.org/10.1007/BF03326046

114. Neppolian B, Ashokkumar M, Sáez V et al (2012) Hybrid sonochemical treatments of wastewater: sonophotochemical and sonoelectrochemical approaches. Part II: sonophoto-catalytic and sonoelectrochemical degradation of organic pollutants. In: Sharma SK, Sanghi R (eds) Advances water treatment pollution prevention. Springer, Dordrecht, pp 303–336

115. Wu TY, Guo N, Teh CY, Hay JXW (2013) Advances in ultrasound technology for environmental remediation. In: Wu TY, Guo N, Teh CY, Hay JXW (eds) Advances ultrasound technology environment remediation. Springer, Dordrecht, pp 13–93

116. Anjaneyulu Y, Sreedhara Chary N, Samuel Suman Raj D (2005) Decolourization of industrial effluents—available methods and emerging technologies—a review. Rev Environ Sci Biotechnol 4:245–273. https://doi.org/10.1007/s11157-005-1246-z

117. Onat TA, Gümüşdere HT, Güvenç A, et al (2010) Decolorization of textile azo dyes by ultrasonication and microbial removal. Desalination 255:154–158. http://doi.org/10.1016/j.desal.2009.12.030

118. Sathian S, Rajasimman M, Radha G et al (2014) Performance of SBR for the treatment of textile dye wastewater: optimization and kinetic studies. Alexandria Eng J 53:417–426. https://doi.org/10.1016/j.aej.2014.03.003

119. Lin H, Chen J, Wang F et al (2011) Feasibility evaluation of submerged anaerobic membrane bioreactor for municipal secondary wastewater treatment. Desalination 280:120–126. https://doi.org/10.1016/j.desal.2011.06.058

120. Martin-Garcia I, Monsalvo V, Pidou M et al (2011) Impact of membrane configuration on fouling in anaerobic membrane bioreactors. J Memb Sci 382:41–49. https://doi.org/10.1016/j.memsci.2011.07.042

121. Xing CH, Tardieu E, Qian Y, and Wen X (2000) Ultrafiltration membrane bioreactor for urban wastewater reclamation. J Memb Sci 177:73–82. doi:https://doi.org/10.1016/S0376-7388(00)00452-X

122. Jegatheesan V, Pramanik BK, Chen J et al (2016) Treatment of textile wastewater with membrane bioreactor: a critical review. Bioresour Technol 204:202–212. https://doi.org/10.1016/j.biortech.2016.01.006

123. Kim HW, Oh HS, Kim SR et al (2013) Microbial population dynamics and proteomics in membrane bioreactors with enzymatic quorum quenching. Appl Microbiol Biotechnol 97:4665–4675. https://doi.org/10.1007/s00253-012-4272-0

124. Wang E, Zheng Q, Xu S, Li D (2011) Treatment of methyl orange by photocatalysis floating bed. Procedia Environ Sci 10:1136–1140. https://doi.org/10.1016/j.proenv.2011.09.181

125. Chakrabarti S, Dutta BK (2004) Photocatalytic degradation of model textile dyes in wastewater using ZnO as semiconductor catalyst. J Hazard Mater 112:269–278. https://doi.org/10.1016/j.jhazmat.2004.05.013

126. Paschoal FMM, Anderson MA, Zanoni MVB (2009) The photoelectrocatalytic oxidative treatment of textile wastewater containing disperse dyes. Desalination 249:1350–1355. https://doi.org/10.1016/j.desal.2009.06.024

127. Arslan I, Balcioglu IA, Banheman DW (2000) Heterogeneous photocatalytic treatment of simulated dyehouse effluents using novel TiO$_2$ photocatalyst. Appl Catal B Environ 26:193–206. https://doi.org/10.1016/S0926-3373(00)00117-X

128. Hong J, Otaki M (2003) Effects of photocatalysis on biological decolorization reactor and biological activity of isolated photosynthetic bacteria. J Biosci Bioeng 96:298–303. https://doi.org/10.1016/S1389-1723(03)80197-4

129. Carneiro PA, Osugi ME, Sene JJ et al (2004) Evaluation of color removal and degradation of a reactive textile azo dye on nanoporous TiO_2 thin-film electrodes. Electrochim Acta 49:3807–3820. https://doi.org/10.1016/j.electacta.2003.12.057

130. Paz A, Carballo J, Pérez MJ, Domínguez JM (2017) Biological treatment of model dyes and textile wastewaters. Chemosphere 181:168–177. https://doi.org/10.1016/j.chemosphere.2017.04.046

131. Shen J, Smith E (2015) Enzymatic treatments for sustainable textile processing. In: Blackburn RS (ed) Sustainable apparel production processing recycling. Woodhead Publishing, Sawston, pp 119–133

132. Madhu A, Chakraborty JN (2017) Developments in application of enzymes for textile processing. J Clean Prod 145:114–133. https://doi.org/10.1016/j.jclepro.2017.01.013

133. Peralta-Zamora P, Kunz A, De Moraes SG et al (1998) Degradation of reactive dyes I. A comparative study of ozonation, enzymic and photochemical processes. Chemosphere 38:835–852. https://doi.org/10.1016/S0045-6535(98)00227-6

134. Sahinkaya E, Yurtsever A, Çınar Ö (2017) Treatment of textile industry wastewater using dynamic membrane bioreactor: impact of intermittent aeration on process performance. Sep Purif Technol 174:445–454. https://doi.org/10.1016/j.seppur.2016.10.049

135. Friha I, Bradai M, Johnson D et al (2015) Treatment of textile wastewater by submerged membrane bioreactor: in vitro bioassays for the assessment of stress response elicited by raw and reclaimed wastewater. J Environ Manage 160:184–192. https://doi.org/10.1016/j.jenvman.2015.06.008

136. Tobien T, Cooper WJ, Nickelsen MG et al (2000) Odor control in wastewater treatment: the removal of Thioanisole from water a model case study by pulse radiolysis and electron beam treatment. Environ Sci Technol 34:1286–1291. https://doi.org/10.1021/es990692v

137. Westerhoff P, Mezyk SP, Cooper WJ, Minakata D (2007) Electron pulse radiolysis determination of hydroxyl radical rate constants with Suwannee river fulvic acid and other dissolved organic matter isolates. Environ Sci Technol 41:4640–4646. https://doi.org/10.1021/es062529n

138. Harrelkas F, Paulo A, Alves MM et al (2008) Photocatalytic and combined anaerobic-photocatalytic treatment of textile dyes. Chemosphere 72:1816–1822. https://doi.org/10.1016/j.chemosphere.2008.05.026

139. Seshadri S, Bishop PL, Agha AM (1994) Anaerobic/aerobic treatment of selected Azo dyes in wastewater. Waste Manag 14:127–137. https://doi.org/10.1016/0956-053X(94)90005-1

140. Minke R, Rott U (1999) Anaerobic treatment of split flow wastewater and concentrates from the textile processing industry. Water Sci Technol 40:169–176. https://doi.org/10.1016/S0273-1223(99)00377-7

141. Lacasse K, Baumann W (2004) Environmental considerations for textile processes and chemicals. In: Lacasse K, Baumann W (eds) Textile chemicals. Springer, Berlin, pp 484–647

142. Siddique K, Rizwan M, Shahid MJ et al (2017) Textile wastewater treatment options: a critical review. In: Anjum NA, Gill SS, Tuteja N (eds) Enhancing cleanup environmental pollutants. Springer International Publishing, Cham, pp 183–207

143. Amuda OS, Deng A, Alade AO, Hung Y-T (2008) Conversion of sewage sludge to biosolids. In: Wang LK, Shammas NK, Hung Y-T (eds) Biosolids engineering management. Humana Press, Totowa, pp 65–119

144. Yadav A, Garg VK (2011) Industrial wastes and sludges management by vermicomposting. Rev Environ Sci Biotechnol 10:243–276. https://doi.org/10.1007/s11157-011-9242-y

145. Karn SK, Kumar A (2015) Hydrolytic enzyme protease in sludge: recovery and its application. Biotechnol Bioprocess Eng 20:652–661. https://doi.org/10.1007/s12257-015-0161-6

146. He L, Du P, Chen Y et al (2017) Advances in microbial fuel cells for wastewater treatment. Renew Sustain Energy Rev 71:388–403. https://doi.org/10.1016/j.rser.2016.12.069

147. Zhang M, Yang C, Jing Y, Li J (2016) Effect of energy grass on methane production and heavy metal fractionation during anaerobic digestion of sewage sludge. Waste Manag 58:316–323. https://doi.org/10.1016/j.wasman.2016.09.040

148. Yang S, Hai FI, Price WE et al (2016) Occurrence of trace organic contaminants in wastewater sludge and their removals by anaerobic digestion. Bioresour Technol 210:153–159. https://doi.org/10.1016/j.biortech.2015.12.080

149. Patinvoh RJ, Osadolor OA, Sárvári Horváth I, Taherzadeh MJ (2017) Cost effective dry anaerobic digestion in textile bioreactors: experimental and economic evaluation. Bioresour Technol 245:549–559. https://doi.org/10.1016/j.biortech.2017.08.081

150. Bahar S, Ciggin AS (2016) A simple kinetic modeling approach for aerobic stabilization of real waste activated sludge. Chem Eng J 303:194–201. https://doi.org/10.1016/j.cej.2016.05.149

151. Sonai GG, de Souza SMAGU, de Oliveira D, de Souza AAU (2016) The application of textile sludge adsorbents for the removal of Reactive Red 2 dye. J Environ Manage 168:149–156. https://doi.org/10.1016/j.jenvman.2015.12.003

152. Volmajer Valh J, Majcen Le Marechal A, Vajnhandl S et al (2011) Water in the textile industry. In: Wilderer WS (ed) Treatise water science. Elsevier, Oxford, pp 685–706

153. Velghe I, Carleer R, Yperman J, Schreurs S (2013) Bioresource technology study of the pyrolysis of sludge and sludge/disposal filter cake mix for the production of value added products. Bioresour Technol 134:1–9. https://doi.org/10.1016/j.biortech.2013.02.030

154. Faubert P, Barnabé S, Bouchard S et al (2016) Pulp and paper mill sludge management practices: what are the challenges to assess the impacts on greenhouse gas emissions? Resour Conserv Recycl 108:107–133. https://doi.org/10.1016/j.resconrec.2016.01.007

155. Ahmadi M, Jorfi S, Kujlu R et al (2017) A novel salt-tolerant bacterial consortium for biodegradation of saline and recalcitrant petrochemical wastewater. J Environ Manage 191:198–208. https://doi.org/10.1016/j.jenvman.2017.01.010

156. Tomei MC, Mosca Angelucci D, Daugulis AJ (2016) Sequential anaerobic-aerobic decolourization of a real textile wastewater in a two-phase partitioning bioreactor. Sci Total Environ 573:585–593. https://doi.org/10.1016/j.scitotenv.2016.08.140

New Tools and Techniques for Measuring Sustainability in Clothing

P. Senthil Kumar and P. R. Yaashikaa

Abstract Textile industry is one of the fundamental ventures which fulfills one of the essential needs of individuals and subsequently moves toward becoming as an unavoidable piece of human's life. Measuring sustainability in textile industry is a complex examination technique. It is directed for supporting basic leadership and strategy in an expansive ecological, monetary, and social setting. The research on sustainability measurement flourishes with various tools, techniques, models, and systems, and the measure of writing on sustainability evaluation connected to the field of assembling is quickly developing. Environmental, social, and monetary effects are significant mainstays of sustainability in textile industry. Measuring sustainability has been viewed as troublesome because of the subjective idea of the deliberate issues. Higg Index has turned out to be a far-reaching self-measuring tool for associations in the clothing sector to utilize. Picking a correct tool to suit the requirements of various players in the field of sustainability is particularly pivotal. Sustainability can be evaluated for a product, venture, and furthermore for an area and a nation. Measuring tools are characterized by different ways. Not all sustainability measuring tools can be utilized for textile network, and a portion of the tools are very apparent to survey the sustainability of textile outcomes.

Keywords Sustainability · Tools and techniques · Environmental
Economic and social supportability · Lifecycle assessment · Sustainable Apparel
Coalition · Higg Index

P. Senthil Kumar (✉) · P. R. Yaashikaa
Department of Chemical Engineering, SSN College of Engineering,
Chennai 603 110, India
e-mail: senthilchem8582@gmail.com; senthilkumarp@ssn.edu.in

© Springer Nature Singapore Pte Ltd. 2018
S. S. Muthu (ed.), *Sustainable Innovations in Apparel Production*,
Textile Science and Clothing Technology,
https://doi.org/10.1007/978-981-10-8591-8_3

1 Introduction

Textile industries assume a critical part in the financial advancement of nation. Sustainability in clothing will require a total change in the originator, makers, advertisers, and customers. Consumers particularly require a soul for sustainable design operations. A noteworthy driver of human effect on earth frameworks is the demolition of biophysical assets, and particularly the earth's environments. The ecological effect of a group or of mankind in general depends both on populace and effect per individual, which thus depends in complex routes on what resources are being utilized, regardless of whether those resources are sustainable, and the size of the human movement in respect to the conveying limit of the biological systems included [4, 14]. Proper resource administration can be connected at many scales, from financial segments like horticulture, assembling, and industry, to work associations, the utilization examples of family and individual people, and to the asset requests of individual merchandise and enterprises. The textile business today is a worldwide industry and has vast affects on our environmental condition and in addition on individuals. Many developing elements are considered which are recognized moral from conventional design including utilization of vitality effective procedures, elective vitality, and low-effect colors in assembling [5]. Sustainability measurement is one of the fundamental parts of manageable advancement; i.e., on the off chance that one cannot gauge the level of sustainability on organization level, one does not know whether one does the correct things and is heading the correct way with our change activities. Any manufacturing organization that needs to enhance sustainability faces the test of surveying its execution. Initially, a significant number of the methodologies are created on national or more nearby level, and few are focusing on the level required for application inside assembling. Besides, sustainability appraisal devices may be considered excessively hypothetical and general, or excessively specialized and confused for manufacturing organizations [1]. This makes a correspondence hole between the scholarly and assembling areas. While specialists work to build up the ideal sustainability appraisal tools, producers do not know how or from where to begin their voyage toward more feasible practices. Thirdly, the larger part of existing evaluation tools is gone for outer detailing and needs helpful data for inward chiefs. Textile industry comprises of a very intricate and a monstrous store network. There are many procedures engaged with getting clothing item at last made, and there are many accomplices assuming a key part in this protracted store network to create an apparel item. This industry is such a broadened one with numerous strands and their committed procedure lines, and even with a similar procedure line, there are numerous variations in the procedures [11, 15]. Maintainability parts of materials must investigate the whole store network of materials and the garments with the scales to evaluate and further to enhance the social, ecological, and financial effects at each phase of the whole inventory network. Makers and retailers have been centering their endeavors in diminishing the ecological effects of materials in all their lifecycle stages and furthermore enhancing the social angles.

2   Sustainable Development

Sustainable development is an advancement that addresses the issues of the present without trading off the capacity of future ages to address their own issues. Practical advancement is the mission of the general public, and it is firmly connected to the utilization of innovation. Without innovation change, society would have danger of crumple. To keep up adjust between innovation for enhancing society and innovation for supportability has now turned into an advantage. Sustainable development is a specific kind of advancement that is portrayed by specific criteria. These criteria of maintainability and transformative advancement can be determined clearly [6]. They give specific introductions to the frameworks and on-screen characters that are a piece of the advancement, making them inclines toward specific activities, ways and effects contrasted and others. For evaluating progress and genuine or expected outcomes of activities, the on-screen characters require an extensive arrangement of pointers depicting the condition of the frameworks under their care and of their condition. Material procedures and items, obligation of innovation engineers and architects are much more unavoidable. Material and apparel industry is a standout among the most extended segments on the planet. Each and every individual takes an interest in some piece of the material division in the generation, assembling, or buyer parts [9]. Material assembling chain begins with fiber developing of regular fiber and assembling of man-made fiber. Utilization of fiber-based items has been expanding because of populace development of developing markets, rising life models, age structure in the creating nations, an expansion of material-based item composes, and measure of their utilization. Assembling advancements that are used at the generation periods of a material thing vary depending on the item properties.

3   Sustainability Issues in Textile Process

Textile dyeing includes the utilization of various distinctive chemicals and helpers to help the coloring procedure. A few helpers (e.g., scattering specialists, supports, dedusting operators) are as of now contained in the dyestuff definition, yet other assistant chemicals are additionally added amid handling to help the readiness, shading, and after washing forms. Since assistants as a rule do not stay on the substrate subsequent to coloring, they are eventually found in the discharges from a dyehouse. Earth mindful color application grasps the settled '3R' standards of contamination counteractive action—i.e., reduce, re-utilize, reuse—and the best contamination counteractive action rehearse for material wet preparing is 'correct first-time' coloring. Restorative measures, for example, shading options or in the most pessimistic scenario strip and re-color are largely substance, vitality, and water escalated and add significantly to the contamination stack. The accompanying are couple of parts for enhancing the supportability in material process particularly coloring. Accurate color communication

- Waste minimization and contamination control;
- Intelligent process choice for enhanced asset efficiency;
- Intelligent color choice for concoction consistence;
- Intelligent color choice for item toughness;
- In spite of the current financial challenges, it is hard to perceive how the material business would ever turn the clock back and disregard issues of contamination, item security, and asset preservation.

Proposals for enhancing supportability

Some developing patterns, as the business endeavors toward a more practical plan of action for material and dress produce and utilization, are exhibited beneath:

- Low vitality coloring forms (conceivable compound actuated, or biomimetic frameworks);
- Low water coloring and printing forms (advancement of froth procedures, ink fly printing and coloring, and reconsideration of dissolvable-based procedures with full dissolvable recuperation);
- A restored accentuation on right-first-time coloring;
- More noteworthy accentuation on item toughness and a move far from 'quick form';
- Expanding prescriptive specification of colors and chemicals by retailers/brands to guarantee accomplishment of execution specification with respect to metamerism, speed, and RSL consistence;
- Expanding straightforwardness and traceability in supply chains;
- Advancement of important measurements to evaluate supportability of supply chains, e.g., carbon footprinting, water footprinting, and contamination stack/unit of creation;
- Expanded cooperation between performing artists along the entire production network from homestead to store;
- Nearer cooperation between retailer/marks on production network gauges, ecological prerequisites, and synthetic confinements;
- More noteworthy consciousness of supportability in mold and material plan.

4   Selection of Tool for Measuring Sustainability

A key segment of supportability evaluations is the examination of various task/approach options. Appraisal apparatuses are characterized as the different diagnostic systems that can be utilized to encourage the examinations. After right around 25 years of level-headed discussion, there is no deficiency of maintainability appraisal devices. As indicated by their presumptions and their valuation viewpoint, maintainability evaluation instruments can be separated into three general classifications: fiscal, biophysical, and pointer-based. What is extremely missing, in any case, are rules and criteria on the best way to pick between these apparatuses [8]. The determination of evaluation instruments is typically performed by the analyst(s) and for the most part relies upon time/information/budgetary requirements, the capabilities of experts, and the scope of devices available to them as opposed to on a strong hypothetical premise or the setting of the general appraisal.

4.1 Tool Selection Theory

The presumptions made by each device class are much of the time profoundly esteem loaded. Basically, these presumptions direct the accompanying:

(a) The valuation viewpoint, of the general appraisal;
(b) The reception of a reductionist or a non-reductionist point of view amid the appraisal;
(c) The agreeableness of exchange offs between the distinctive maintainability issues.

Financial instruments are inclination-based. They depend on models of human conduct and lay on the suspicion that esteem emerges from the subjective inclinations of people. Neoclassical financial valuation instruments basically catch a man's eagerness to pay (WTP) for devouring a product/benefit or the readiness to acknowledge (WTA) remuneration for relinquishing this utilization. Biophysical instruments measure the measure of regular assets that have been contributed amid the generation of a decent or an administration. Biophysical instruments allot esteem in the light of the characteristic properties of items by measuring basic physical parameters and after that making an interpretation of them into a shared factor or unit of estimation [3]. The biophysical esteem measured is viewed as an intermediary to natural effect. This implies the best task/arrangement elective is the one that outcomes in the assignment of the least measure of common assets. Marker-based instruments likewise involve a progression of exceptionally esteem loaded methodological decisions, especially amid pointer choice, measuring, standardization, and conglomeration. Such methodological decisions basically manage whether a particular valuation point of view is received amid the appraisal and as an expansion the part that people 'accept' inside the evaluation. Moreover, devices, for example CBA and CIs, that contain express accumulation steps basically permit exchange offs between the distinctive manageability issues (commensurability of supportability issues) receiving subsequently a feeble maintainability point of view.

4.2 Outline of Surviving Sustainability Measurement Applications

Past investigations demonstrate that there is a wide range of methodologies toward the improvement of supportability evaluation apparatuses. A gathering of analysts dissected eight supportability appraisals and distinguished that some of these apparatuses concentrate on item evaluation as opposed to assembling forms. Different instruments address just natural viewpoint, though the rest of being thought about excessively broad for functional application. A few instruments expecting to direct reasonable assembling examination give only an ecological appraisal. Illustrations incorporate, among others, an action-based, protest arranged technique for maintainability appraisal of fabricated parts amid assembling period of life, manageability appraisal of assembling stream for another segment, producing ecological execution assessment, supportability evaluation in the light of a blended whole number direct programming model. As of late, noteworthy endeavors have been made to create maintainability appraisal instruments tending to each of the three parts of manageability. In any case, the greater part of the devices is created for particular items, procedures, or parts of assembling framework. Cases of such devices include: manageability appraisal of assembling framework reuse in view of analytic hierarchy process (AHP); assessment of item in light of fluffy thorough evaluation technique; maintainability execution estimation for item life cycle; coordinating economical assembling appraisal for a generation work cell in light of AHP and multi-criteria basic leadership strategy; a grid assessment demonstrate for supportability evaluation of assembling advances; maintainability appraisal structure for remanufacturing industry; and Product Sustainability Assessment (PROSA).

5 Environmental Sustainability

The material business makes huge condition impacts for the duration of the life cycle of material items. The collected hurtful substance residuals utilized as a part of material creation are discharged; regularly untreated and straightforwardly into water sources will in the long run pulverize the dirt, water, and the earth. Together with the development of crude materials, in the material business, the destructive impacts are made for the duration of the life cycle of material items. Notwithstanding concoction release into water sources, amid material assembling forms, for example, coloring, printing, and completing, an over the top measure of water, petroleum derivatives, and electrical vitality is expended [16]. Because of serious ecological effects made by the material business, natural manageability has turned into a key worry for material makers' organizations and buyers' ways of life and item buy decisions.

Ecological manageability is a business system for utilizing forms without producing hurtful impacts to the earth and common assets for the duration of the life cycle (e.g., comprising of accumulation, handling, application, renewal, utilization, and transfer). Thus, in the material business, the utilization of vitality, chemicals, and water is major ecological effect generators for the duration of the life cycle of items. To guarantee ecological maintainability, attire originators ought to make items in view of naturally and socially capable outline methodologies and patterns; the production network must think about its effects on society, economy, and condition for their business rehearses. There are some material organizations that have set ecological maintainability as a need in their business hones. The International Organization for Standardization (ISO) discharged ISO (ecological administration—lifecycle appraisal), the enhanced lifecycle evaluation (LCA) principles intended to survey the natural effect of an item for the duration of its life cycle from crude materials to assembling and from utilization to disposing of utilized items. In light of ISO norms, in the material business, there are five stages in an item's life cycle to be considered for natural supportability: (a) material stage, (b) fabricating stage, (c) retail stages, (d) utilization stages, and (e) transfer stages.

6 Economic Sustainability

'Economic/financial sustainability' alludes to an arrangement of generation that fulfills display utilization levels without trading off future needs. On a framework level, a financially practical framework must have the capacity to create merchandise and ventures on a proceeding with premise, to keep up sensible levels of government and outside obligation and to maintain a strategic distance from outrageous sectoral lopsided characteristics which harm farming or modern generation. Manageable clothing will give another market to extra openings for work, persistent net stream of cash in the economy, and the lessening of crude materials and virgin assets, however as far as which common assets the creation procedure may draw upon, how assets are utilized and renewed, the general effect of the last generation on nature, and where the item winds up following its transfer [13]. There is a great deal of development made being used of these filaments in items, e.g., attire, articles of clothing and home material and to make this item proficiently new and imaginative advances and generation strategies created. Research proceeds in material division and the utilization of regular fiber increments when contrasted with synthetic filaments.

6.1 Financial Issues for Enhancing Maintainability

Sourcing reasonable creation can be performed by choosing makers that are known to work in a socially capable way. Neighborhood creation is a significant resource for enhancing item manageability. This will likewise enhance the monetary maintainability of an item. To lessen squander, the eco-accommodating coloring strategy like decreased water coloring or waterless coloring technique for the texture handling. An on item mark with guidelines to reuse, reuse of the item, in this manner diminishing waste, ought to be added to the name. Clean creation can be accomplished by choosing producers that make a pledge to the utilization of option or sustainable power sources. Common filaments are regularly seen as the most ecologically ridiculously from inexhaustible sources and utilize less vitality and chemicals to create than fabricated strands [17]. Eco-accommodating filaments, creation strategies that request less normal assets and utilize nature to their advantage ought to be utilized. The originators, makers, and merchandiser ought to present normal and financial naturally low-effect colors in materials preparing to better item supportability. In any case, so as to make a sparing feasible material industry one has to reexamine the way one plans, creates, expends, and discards design and one has to change form, in a manner of speaking. The answer for the issue lies in understanding the genuine expenses of delivering materials items.

7 Social Sustainability

Social supportability is the limit of a gathering of individuals to develop setups to meet the necessities of its current constituent and furthermore to help the capacity of future ages to maintain a solid group. The technique for moving toward manageability and its advancement is otherwise called social supportability. The term social supportability has less mindfulness among individuals when contrasted with financial and ecological maintainability. The basic importance of social supportability is the limit of a societal framework, similar to a nation, to perform at an unmistakable level of social prosperity inconclusively. Supportability ternion contains ecological maintainability, financial manageability, and social supportability. It is generally called as exchange of financial, social, and natural variables of progression. The possibility of 'social supportability' includes social esteem, liveability, prosperity esteem, gather change, social capital, social reinforce, human rights, work rights, put making, social commitment, social value, social expertise, aggregate flexibility, and human modification [10]. Maintainability is the powerful meeting of present social, financial, and natural necessities without exchanging off the limit of future period to address their own issues, got from the most generally perceived significance of supportability. Maintainability is organizing human flourishing with acknowledged dependability. Social commitment consolidates supportability with particular issues, for example resource usage, defilement, buyer thriving, human rights, prosperity and security, thing moderateness, and quality. The significant perils in material industry are compound, mechanical, and natural. The organization or the association needs to routinely check and report the national laws and directions concerning working condition security. The organization should then develop a tradition through which to realize these laws.

8 Sustainability Measurement Tools

Sustainability assessment tools can be separated into three classifications: product/item-related evaluation, project/venture-related evaluation, and sector/segment- and nation-related appraisal. Moreover, indicators or indices are grouped. The reason for such classifying is to characterize which of the three parts of sustainable improvement (economic/monetary, environment condition, and social) are satisfied by various evaluation tools, and at which level maintainability appraisal operations are to be actualized. Any assessment tool for sustainability must focus on three elements, namely environmental sustainability, economic sustainability, and social sustainability. There are many dedicated tools, and methods were developed for assessing each element of sustainability. One of the major areas which needs to be discussed when it comes to sustainability assessment is lifecycle assessment (LCA) or lifecycle thinking. The indicators are one of the essential tools captured by all the three classification methods. Figure 1 shows the elements of sustainable system.

8.1 Carbon Footprint

The fundamental supporter of a worldwide temperature alteration is carbon dioxide, which represents about 80 for every penny of emanations from the industrialized nations. The carbon impression uncovers the amount CO_2 altogether is radiated along the esteem chain of an item. It is the aggregate arrangement of ozone-harming substance (GHG) discharges caused by an association, occasion, or item. It is computed for the day and age of a year and communicated as far as the measure of carbon dioxide, or its likeness different GHGs transmitted. As nursery, gases delivered by human exercises gather and their fixation increments in the environment cause an earth-wide temperature boost. Carbon footprinting is a key natural bookkeeping device for business chiefs, strategy creators, and non-legislative associations endeavoring to recognize relief measures that diminish the risk of environmental change [12]. The material business is progressively occupied with carbon footprinting as a piece of arrangement advancement and item outline. Carbon footprinting is a simplified type of item ecological impression

Fig. 1 Elements of sustainable system

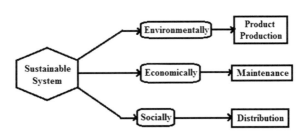

(PEF) figuring, and the two devices are at last in light of ISO, the worldwide standard forever cycle evaluation (LCA). A key component of the first venture in carbon footprinting is definition of the 'useful unit'- a quantitative depiction of the benefit an item is required to give which empowers two items to be looked at on a reasonable premise. Stock investigation is the most tedious advance in carbon footprinting. The third step of carbon footprinting, affect evaluation, is the point at which the contrasts between the nursery gasses are considered. Carbon footprinting normally considers all the nursery gases, for example carbon dioxide, methane, nitrous oxide, hydrofluorocarbons, perfluorocarbons, and sulfur hexafluoride.

The understanding period of a carbon impression ponder is an opportunity to formally reflect on the significance of results of the first three stages and most likely to do parts of them once more. Once in a while, just carbon-containing nursery gases (CO_2 and methane) or in some cases even just CO_2 have been incorporated when the carbon impression is ascertained. A key test in evaluating the carbon impression (as in LCA all in all) is to decide the extent of the framework to be surveyed that is the thing that exercises ought to be incorporated into the appraisal. Carbon footprinting can be connected at many scales from singular items to organizations and national enterprises. The philosophies and guidelines significant to an investigation rely on the size of the examination and its motivation. There are two major strategies being used for acquiring stock information for carbon foot-printing: process examination and information yield investigation. Process investigation is an approach made well known in the field of LCA. It regularly includes measuring physical information sources and yields, and any vitality flows at the level of an item, process, or business. Info yield examination has truly been utilized at the local or national level. Process examination has the benefit of specificity; however, the drawback is that it is moderately tedious. Info yield examination has the preferred standpoint that it requires little exertion for the expert to incorporate information for little flows, and the danger of truncation mistake (i.e., lost frame-work parts) is decreased. It has the detriment that by utilizing normal emanations for each portion of the economy, the natural execution of the most noticeably bad polluters in a section is belittled and ecological pioneers are rebuffed with overestimates.

The evaluation of carbon footprint (CFP) is a vital tool and a reason for over-seeing and controlling ozone-harming substance emanations. At the product level, the CFP and carbon name could add to the low-carbon utilization mode by giving more carbon data to customers, in this way assuming a vital part in inciting society toward a low-carbon mode. China is the biggest textile and piece of clothing producer and customer, on the planet. Research on the CFP of materials is sig-nificant in the administration of residential ozone-depleting substance outflows, and in conveying carbon data and completing important transactions in worldwide exchange. A near investigation of the consequences of various kinds of fabric demonstrates that the industrial CFP of wool textures is very nearly three times that of cotton textures. The industrial CFP of yarn-colored texture is higher than that of colored texture, by 70.8%, and the modern carbon impression of texture made by the plain weave process is higher than texture made by the rib procedure, by 76.2%.

Indirect industrial CFP, the major source of power utilization, contributes around 87% of the aggregate industrial CFP, while coordinate or direct industrial CFP just contributes around 13%. The utilization of energy, for example power, steam, and coal, is the primary source of industrial CFP [18].

8.1.1 Methods to Reduce Carbon Footprint

Ecological maintainability can be accomplished by taking a gander at the full life cycle of our attire, from the outline and materials sourcing process onward. An individual, country, or association's carbon impression can be measured by embracing a GHG outflows appraisal. Once the span of a carbon impression is known, a technique can be formulated to lessen it, e.g., by mechanical improvements, better process and item administration, changed Green Public or Private Procurement (GPP), Carbon Catch, utilization techniques. The moderation of carbon impressions through the advancement of option ventures, for example, sun powered or wind vitality, or reforestation, speaks to one method for lessening a carbon impression and is regularly known as carbon offsetting. To make new green worldview, the material and attire industry needs to receive 3R concept, i.e., diminish, reuse, and recycle.

Reduce: Low-carbon impression forms cut expenses by diminishing misuse of crude materials and vitality. Water and vitality utilization decreases by the material coloring, and complete area can help lessen worldwide carbon dioxide outflows. By sparing vitality and water, the material business can spare a great deal of cash, as well as help to back off environmental change.

Reuse: Effluents of artificially treated materials are released in water. Treatment of wastewater got from synthetically treated materials is an unquestionable requirement. Utilization of chrome severe coloring and restricting the discharge of copper, chromium, and nickel into water decreases polluting influences in colors and shades. Utilizing coloring transporters with high chlorine substance ought to be sidestepped. Amid the way toward blanching, elective operators that are less or not dangerous can be utilized.

Recycle: The material and attire industry should more use reused filaments. An extensive variety of creative, practical attire can be produced using reused materials. We should deal with the approaches to battle 'quick form' and to diminish its negative natural effect as the issues of material reusing, shabby garments, or 'disposable mold' influences every one of us.

Advantages of carbon footprint

By measuring and deciding the sources of ozone-depleting substance release of an item amid the generation chain, it is conceivable to set up targets, systems, and activities to diminish these discharges, and in this manner, the creation of products is more viable and monetary. Another preferred advantage of the CFP is to give information to consumers and bring issues to light with the goal that it can lessen its own effect on a worldwide temperature alteration.

Disadvantage

- Climate change;
- Release of greenhouse gases and global warming;
- Excess utilization of natural resources resulting in depletion.

8.2   Water Footprint

Water footprint is also sometimes called the CFP or impression. The distinction is that it is measured by water utilization, similar to the name infers. The Water Footprint Network isolates the water in three distinct classifications in view of where it originates from. These classes are blue water, green water, and dark water. Blue water alludes to the crisp surface and groundwater; green water is the precipitation that is put away in the dirt or over vegetation; and dim or gray water is the water that gets dirtied in the assembling procedure. The textile business is the third biggest buyer of water on the planet. It utilizes tremendous measure of water all through handling operations like coloring, completing, texture arrangement steps, including desizing, scouring, fading, and mercerizing [2]. Besides water utilization, the effect of textile wastewater on water quality must be considered. After every single material process, water for the most part came back to our biological community without treatment—implying that the wastewater contains chemicals, for example formaldehyde, chlorine, and heavy metals like lead and mercury. These chemicals cause both environmental harm and human illness. In spite of the fact that numerous endeavors have been made to diminish or reuse water through zero release frameworks, selection has been eased back because of the failure to reliably imitate exact shading, and cost required for wastewater treatment to meet water quality measures for the coloring forms.

Managing water footprint

- Utilization of water in an efficient way;
- Better water management and conservation techniques;
- Minimizing the loss of blue and gray water;
- Increasing the consumption of green water;
- Water recycling.

Advantages

The quality of the WF idea is that it gives an expansive point of view on the water administration of the framework and takes into account a more profound comprehension of water utilization. The WF incorporates water utilization and contamination over the total supply chain. It gauges water use of various sorts, for example blue, green, and gray waters. It additionally associates water uses to particular places and times and assesses the hydrological supportability of a framework.

Disadvantages

The shortcomings of the WF are that it signifies to only the amount of water utilized without an estimation of the related natural effects, the absence of required information, and the quantification of gray WF is subjective, and no vulnerability thinks about are accessible despite the fact that vulnerability can be critical. There is as yet more prominent potential for development seeing accounting gauges and in addition information scope, disaggregation, and standard.

8.3 Product-Related Assessment

The item-related devices concentrate on the streams regarding generation and utilization of products and enterprises. Based on a comparable stream point of view, they are firmly identified with the local stream markers. In any case, the devices in this class concentrate on assessing diverse streams in connection with different items or administrations rather than locales. They assess asset utilize and ecological effects along the creation chain or through the life cycle of an item. The points of recognizing specific dangers and wasteful aspects to help basic leadership are like the provincial stream markers. These tools do not incorporate nature–society frameworks as they are predominantly concentrating on ecological viewpoints. In any case, lifecycle costing (LCC) tools may incorporate natural and monetary measurements. Item-related instruments permit both review and forthcoming appraisals that help basic leadership. The most settled and very much created instrument in this classification is the LCA. LCA has been utilized as a part of fluctuating structures in the course of recent years to assess the natural effects of an item or an administration for the duration of its life cycle. It is an approach that breaks down genuine and potential weight that an item has on the earth amid crude material securing, generation process, utilization and transfer of the item. LCA comes about providing data to choices with respect to item advancement and eco-plan, generation framework upgrades, and item decision at the purchaser level.

8.4 Lifecycle Assessment

Lifecycle assessment tool in textile industry is utilized for measuring the item, fabricating process as for its effect on condition amid the item life cycle. The LCA tool discovers its application while settling on choices with respect to controlling natural hazards of the item or creation process. This instrument likewise helps in correlation of ecological effects caused by various items. The stages of LCA were shown in Fig. 2.

LCA is an instrument for surveying an item, process, or administration in regard to its ecological effect amid as long as it can remember cycle from the 'support to grave.' It might be utilized, for instance, for item advancement and change, vital arranging, open policymaking, and advertising. The LCA apparatus is utilized for finding the problem areas in the life cycle with a specific end goal to have the capacity to settle on the best choices on limiting the ecological weights of the item, process, or administration. It is additionally utilized for examinations between, for example, unique items with respect to natural effect [7]. The objective expresses the proposed application, the explanation behind completing the examination and the intended interest group. The extension portrays the broadness, the profundity, and the detail of the investigation. It is essential to characterize a useful unit and the framework limits. The information quality prerequisites ought to be properly indicated. Stock investigation goes for deciding streams of material and vitality between the specialized item framework and nature. Information for input and yield streams is gathered for every unit operation and amassed for the entire life cycle. Information streams could be assets; for example, crude materials, vitality or land and yield streams could be emanations to air, water, or land. Effect appraisal goes for assessing the centrality of potential natural effect in the light of the consequence of the lifecycle stock investigation. The translation step implies that conclusions are drawn and that suggestions can be given and the phases of LCA is shown in Fig. 3.

Strengths of LCA

- LCA has likewise been utilized to evaluate minimization of ozone-harming substance (GHG) outflows in other areas of the general public.

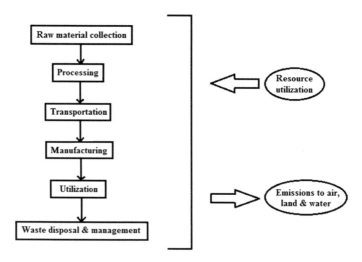

Fig. 2 Lifecycle assessment stages

Fig. 3 Phases of lifecycle
assessment

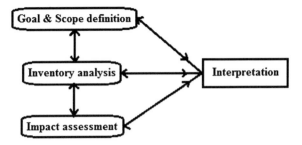

- Measure and oversee parts of the ecological supportability of products, administrations, and operations along the diverse lifecycle stages.
- Distinguish the natural effects of products at all phases of their life cycle.
- Distinguish chances to enhance the natural characteristics at different phases of the product life cycle by executing eco-plan practices to support expanded proficiency, development, and potential cost investment funds.

Limitations

LCA is utilized as a part of a more extensive setting where the requirement for a specific item, process, or administration is addressed. The LCA instrument itself does not contain such contemplations. LCA report does not take social characteristics in the analysis. They do likewise not consider the hazard/opportunity that, if an item progresses toward becoming delivered all the more effective and eco-friendly, customer request may as an outcome increment, causing a natural effect. LCA thinks about portray worldwide or local impacts and not nearby ones. Research is being done on portraying the nearby impacts and fitting them into the LCA structure.

8.4.1 Lifecycle Costing

LCC is a financial approach that aggregates adding up to expenses of an item, process, or movement reduced over its lifetime. The LCC approach is a reasonable talk of convenience for natural basic leadership. On a basic level, LCC is not related to ecological expenses. A conventional LCC is a speculation count that is utilized to rank distinctive venture options to help settle on the best option. There are a wide range of devices for cycle costing. Economic valuation comprises of tools that are not sustainability appraisal systems themselves, yet rather an essential arrangement of devices that can be utilized to help different tools when money-related esteems are required for merchandise and enterprises is not found in the commercial center. Investigation of material and substance streams is likewise utilized for product frameworks. Item vitality investigation measures the vitality that is required to fabricate an item or an administration. It incorporates both immediate and indirect vitality streams.

Calculation of Lifecycle Costing

LCC can be calculated by using two approaches, namely deterministic and probabilistic approaches. In deterministic approach, the discrete values are fixed and are allocated to various factors and any variations are generally ignored. Here the LCC value calculated is fixed. In probabilistic approach, parameters are allocated along with probability distribution. Random numbers are being generated which are used to calculate lifecycle costing. The list of costs involved in calculating lifecycle costing is initial (C_i), installation (C_{ins}), energy (C_e), operational (C_{op}), maintenance (C_m), downtime (C_{dt}), and disposal (C_{dis}) costs.

$$LCC = C_i + C_{ins} + C_e + C_{op} + C_m + C_{dt} + C_{dis}$$

8.5 Project and Country/Sector-Related Assessment

These tools are utilized for supporting choices identified with a strategy or a task in a particular area. Venture-related instruments are utilized for neighborhood scale evaluations, while the approach-related ones concentrate on nearby to worldwide scale appraisals. With regard to sustainability evaluation, incorporated appraisal tools have an ex-risk center and regularly are done as situations. A large number of these coordinated evaluation devices depend on frameworks examination approaches and incorporate nature and society viewpoints. Coordinated evaluation comprises of the wide exhibit of instruments for overseeing complex issues. The setup tools, for example multi-criteria analysis (MCA), risk analysis, vulnerability analysis, and cost–benefit analysis, do not really relate straightforwardly to just sustainability issues and, however, can be stretched out to an assortment of other issue regions crosswise over disciplinary limits.

8.5.1 Multi-criteria Analysis

Multi-criteria decision analysis (MCDA) has been viewed as a reasonable arrangement of techniques to perform sustainability assessments because of its adaptability and the likelihood of encouraging the exchange between partners, experts, and researchers. It has been accounted for that specialist does not as a rule appropriately characterize the explanations behind picking a specific MCDA strategy rather than another. MCA is utilized for measurements in circumstances when there are contending assessment criteria. When all is said in done, MCA distinguishes objectives or targets and after that tries to recognize the exchange offs between them, a definitive objective being identification of the ideal approach. This approach has the benefit of fusing both subjective and quantitative information into the procedure.

8.5.2 Risk Analysis

Risks can be characterized as the same number of things yet at the foundation of each definition is the way that dangers speaks to indeterminate results. These results can be either negative or positive. Basically, hazard administration is the mix of three stages: risk assessment, emanation, and exposure control. Two basic factors that reason issues to Indian material industry are India has an extremely old and divided material mechanical infrastructure and an absolutely lacking and little administration framework for materials. Hazard appraisal is a basic portion in risk administration hypothesis for assessing the potential hazard from a gathering of testing information tending to the wellbeing issues of material and clothing. Hazard evaluation has turned into an expanding prominent and profitable measure in item wellbeing investigation research, and significant enterprises commonly utilize distinctive appraisal strategies. Quantitative and profoundly dependable assessment of material and attire dangers, moderately simple reviewing grouping and straightforwardness in working the assessment procedure are the favorable circumstances that advance the use of hazard appraisal demonstrate in view of fluffy neural system for material and clothing.

8.5.3 Vulnerability Analysis

A vulnerability analysis is the way toward distinguishing, evaluating, and organizing (or positioning) the vulnerabilities in a framework. It might be directed in the political, social, monetary, or ecological fields. Vulnerability evaluation has numerous things in the same manner as hazard appraisal. Evaluations are ordinarily performed by the accompanying advances:

1. Characterizing and grouping resources and abilities (assets) in a framework;
2. Appointing quantifiable esteem and significance to those assets;
3. Distinguishing the vulnerabilities or potential dangers to every asset;
4. Building up a methodology to manage the most genuine potential issues first;
5. Relieving or taking out the most genuine vulnerabilities for the most significant assets.

Vulnerability examinations concentrate both on results for the question itself and on essential and auxiliary outcomes for the encompassing condition. It likewise frets about the potential outcomes of diminishing such results and of enhancing the ability to oversee future episodes.

8.5.4 Cost–Benefit Analysis

Cost–benefit analysis has been built up basically as an apparatus for use by governments in settling on their social and monetary choices. Figure 4 shows the

Fig. 4 Activities in cost–benefit analysis

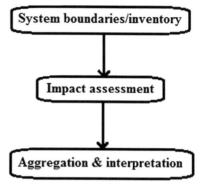

activities in cost–benefit analysis. It gauges expenses and advantages to the group of embracing a specific game plan. It is a basic leadership gadget for assessing exercises that are not valued by the market. It is utilized for assessing open or private venture recommendations by measuring the expenses of the undertaking against the normal advantages. In the domain of maintainability evaluation, CBA can be a successful instrument for measuring the social expenses and advantages of various options regarding, for instance, vitality and transports.

8.6   Indicators

Pointers can give urgent direction to basic leadership in an assortment of ways. They can quantify and align advance toward economical improvement objectives. They can give an early cautioning, sounding the alert so as to counteract financial, social, and ecological harm. Pointers are straightforward measures, frequently quantitative, speaking to a condition of financial, social as well as ecological improvement in a characterized locale—regularly at the national level. Markers ought to contain the accompanying qualities: effortlessness, (a wide) scope, are quantifiable, enable patterns to be resolved, instruments that are touchy to change, and permit auspicious recognizable proof of patterns. Markers and lists, which are ceaselessly measured and figured, take into account the following of longer term supportability patterns from a review perspective. The apparatuses in the class of pointers and files are either non-incorporated, which means they do not coordinate nature–society parameters, or incorporated, which means the instruments total distinctive measurements. There is additionally a subcategory of non-coordinated instruments that spotlights particularly on territorial stream markers, for instance environmental pressure indicators.

  Clothing organizations can utilize key execution markers to distinguish business qualities and shortcomings. Organizations set execution objectives and utilize the quantitative markers to gauge their prosperity. Solid attire organizations utilize enters execution markers in an assortment of business measurements. Showcasing and deals execution pointers enable apparel to organization judge the achievement of estimating methodology and promoting efforts. Since patterns change rapidly, attire organizations should have the capacity to rapidly react to changing item requests. Attire organizations look to accomplish high stock turnover markers. The speedier attire retailers can turn over stock, the greater limit they have for new items and the more improbable it is they have unfashionable, out of date stock close by. Organizations will likewise assess time to market to perceive how rapidly they can make an interpretation of a clothing plan into a sellable item. Clothing organizations that fabricate items ought to routinely assess generation pointers. Standard costing pointers help attire makers comprehend changes underway costs, amount, and quality. Organizations can gauge the sum and cost of the machine hours they used to make their items. They can quantify the normal deformity rate per item to distinguish territories of shortcoming. Price tag differences can enable chiefs to distinguish where less expensive crude materials may bring about lower quality attire items. Attire organizations frequently set target money-related pointers to build their stock cost or bait in new financial specialists. Organizations can gauge their arrival on resources and profit for income to perceive how well they are utilizing the assets accessible to them. Clothing area that are looking for long and here and now financing should concentrate on keeping up low obligation proportions and a high current proportion.

8.7 Sustainable Apparel Coalition Tools

The Sustainable Apparel Coalition (SAC) looks to lead the clothing business toward a mutual vision of manageability based upon a typical approach for assessing maintainability execution. By building up a typical device—the Sustainable Apparel Index—the SAC empowers clothing industry organizations to quantify the natural and social effect of attire generation all through the item life cycle, from configuration to end of utilization or reusing of the item. The motivation behind the SAC is twofold. The part associations will detail intentions to diminish the effect of the attire business with respect to the utilization of water, chemicals, and waste age. Also, the SAC will build up an evaluation apparatus for the estimation of natural and social effects. Its attention is on the institutionalization of the estimation of natural and social effects of clothing and footwear items over the item life cycle all through the esteem chain. The accompanying are few instruments recorded by SAC.

8.7.1 Higg Index

The material business has its own particular industry—particular appraisal instrument called the Higg Index. The Higg Index has been produced by the SAC that is an association framed by a portion of the biggest brands and makers in the material business. The point of the SAC was additionally to build up an apparatus that would serve similarly all associations in the business regardless of on the off chance that they are a brand, producer, or retailer. The Higg Index focuses absolutely on non-money-related pointers to demonstrate the present condition of their social and natural endeavors. The apparatus is moderately new, and huge numbers of its modules are still in beta stage. In this manner, there is an exceptionally predetermined number of studies led on the instrument or its utilization. The oddity of the Higg Index additionally prompts the way that it is at present under development and further improvement. A scorecard-adjusted approach strategy was produced to conquer the issues of Higg Index. Higg Index involved three unique modules: brand, office, and item. Each of these modules can be utilized freely. Late occasions have uncovered a fundamental disappointment in the attire and footwear industry that calls for important outcomes at a foundational level, and the point of the SAC and the Higg Index is to address that need. As per SAC, the goal of the Higg Index is to demonstrate associations their qualities and shortcomings, exhibit open doors for cost reserve funds and development, and catalyze maintainability instruction and joint effort. The Higg Index can be viewed as the SAC's approach to identify with the social and planetary limits introduced in the past part. The Higg Index tends to the effects of the material and design industry in five unique regions: water, biodiversity, vitality, waste, and social. The Higg Index is a continually advancing apparatus, and the SAC recognizes the requirements for enhancements.

8.7.2   Facility Tools

Makers utilize the Higg Index modules to gauge the social and natural execution of their offices. These modules measure impacts at singular processing plants, not the parent organization in general. The Higg Facility devices make open doors for open discussion among store network accomplices, so organizations at each level in the esteem chain altogether perform better. The office apparatuses incorporate Facility Module: Environment—Apparel/Footwear evaluates the execution of materials, bundling and assembling offices and Facility Module for Social/Labor—Apparel/ Footwear beta is utilized for the social and work execution of materials, bundling and assembling offices.

8.7.3   Brand Tools

Clothing, footwear, and materials brands and retailers of all sizes to quantify the ecological, social, and work execution of their plan, sourcing, and operations utilize the Higg Index's Brand Modules. The modules evaluate corporate approaches and practices in each effect region and at each level of supportability, from fundamental, consistence level practices to cutting edge and expansive accepted procedures. There are three brand instruments. Condition: Apparel is utilized to survey attire items with uncommon reference to the specific natural practices at the brand level.

The brand module Environment: Footwear works the same as the past module, yet the item is footwear. The brand module Social/Labor: Apparel/Footwear Beta is utilized to survey the specific social and work rehearses for both clothing and footwear at the brand level.

8.7.4 Product Tools

The Higg Index Product Tools are utilized amid the outline and materials determination phases of creation to survey an item's supportability impacts; they likewise evaluate an item's lifetime manageability affect. Higg Product Tools offer brands and makers data to settle on better decisions at each phase of an item's advancement. The Rapid Design Module (RDM)—Beta and the Material Sustainability Index (MSI) Data Explorer are the two evaluation instruments for measuring the effect of items.

8.7.5 Retail Tools

The Higg Index's Retail Modules are utilized by attire, footwear, and material retailers of all sizes to quantify the natural and social execution of the majority of their operations from marketing to coordinations. These modules evaluate corporate strategies and practices in each effect region and at each level of maintainability, from essential, consistence level practices to cutting edge and expansive prescribed procedures.

9   Conclusion

Textile producing innovations cause some high measure of ecological load on the nature because of their vitality utilization, wastewater release, and different kinds of waste gathering. All assembling operations and advances make unsafe effects on the nature; nonetheless, a large portion of the procedures is fundamental procedures that are customarily acknowledged. There is no probability of zero effect, or totally eradicating of the ecological load, left after processing a textile material. Just conceivable approaches to decrease the natural load are innovation change, proper technology determination, and proficient innovation administration. Current mechanical improvements offer a few chances to make a move of new creation innovation courses and models to assemble supportable material innovation condition. Lifecycle approach to bargain sustainability in material and clothing industry includes ensuring the three parts of sustainability—social, financial, and ecological. Textile and clothing ventures are fundamental parts of the world economy, offering occupations to a huge number. The administration and the business need to cooperate and set up a course of action of movement that finds key issues and recognizes and empties checks to improvement and sourcing approaches. Sustainability evaluation tools contain a few conclusions about what is essential to be measured, how to quantify it, who and in what part should be considered in the appraisal, and what sustainability points of view are both applicable and real. A couple of tools are competent to distinguish particular issues, though none of the current tools bring up conceivable outputs. In this manner, the appropriateness of the current tools is restricted. Taking everything into account, sustainable advancement is to a great extent about individuals, their prosperity, and value in their associations with each other, in a setting where imbalanced nature–society characteristics can influence financial and social dependability.

References

1. Baumgartner S, Quaas M (2010) What is sustainability economics? Ecol Econ 69:445–450
2. Belz F-M, Peattie K (2012) Sustainability marketing: a global perspective, 2nd edn. Wiley, Chichester
3. Binder CR, Feola G, Steinberger JK (2010) Considering the normative, systemic and procedural dimensions in indicator-based sustainability assessments in agriculture. Environ Impact Assess Rev 30:71–81
4. Bostrom M, Micheletti M (2016) Introducing the sustainability challenge of textiles and clothing. J Consum Policy 39:365–375
5. Chen Hsiou-Lien, Burns LD (2006) Environmental analysis of textile products. Cloth Text Res J 24:248–261
6. De Ridder W, Turnpenny J, Nilsson M, von Raggamby A (2007) A framework for tool selection and use in integrated assessment for sustainable development. J Environ Assess Policy Manage 9:423–441
7. E.P.A. (U.S. Environmental Protection Agency) (2006) Life cycle assessment: principles and practice. Tech. Rep. EPA/600/R-06/060, National Risk Management Laboratory
8. Gasparatos A (2010) Embedded value systems in sustainability assessment tools and their implications. J Environ Manage 91:1613–1622
9. Giampietro M, Mayumi K, Bukkens SGF (2001) Multiple-scale integrated assessment of societal metabolism: an analytical tool to measure development and sustainability. Environ Dev Sustain 3:275–307
10. Hasna AM (2008) A review of sustainability assessment methods in engineering. Int J Environ Cult Econ Soc Sustain 5(1):161–176
11. Poveda CA, Lipsett MG (2011) A review of sustainability assessment and sustainability/environmental rating systems and credit weighting tools. J Sustain Dev 4:36–55
12. Sarzynski A (2012) Carbon footprint. Berkshire Encycl Sustain 6:42–45
13. Shen B, Choi TM, Lo KY (2016) Enhancing economic sustainability by markdown money supply contracts in the fashion industry: China vs U.S.A. Sustainability 8, 31
14. Shen B, Li Q, Dong C, Perry P (2017) Sustainability issues in textile and apparel supply chains. Sustainability 9:1592. https://doi.org/10.3390/su9091592
15. Singh RK, Murty HR, Gupta SK, Dikshit AK (2012) An overview of sustainability assessment methodologies. Ecol Indic 15(1):281–299
16. Veleva V, Hart M, Grenier T, Crumbley C (2003) Indicators for measuring environmental sustainability: a case study of the pharmaceutical industry. Benchmarking: Int J 10:107–119

17. Waage SA, Geiser K, Irwin F, Weissman AB, Bertolucci MD, Fish P, Basile G, Cowan S, Cauley H, McPherson A (2005) Fitting together the building blocks for sustainability: a revised model for integrating ecological, social, and financial factors into business decision-making. J Cleaner Prod 13:1145–1163
18. Yan Y, Wang C, Ding D, Zhang Y, Wu G, Wang L, Liu X, Du C, Zhang Y (2016) Industrial carbon footprint of several typical Chinese textile fabrics. Acta Ecilogica Sin 36(3):119–125

Printed in the United States
By Bookmasters